SCIENCE & WONDERS VOLUME III

As X Goes to Infinity

Amy Joy Hess

Ocean to Mountain Publishing
PO BOX 1116
Wallace, ID 83873
www.otmpub.com

This book is based on the true-life experiences of the author, who has taken care to portray the persons and events therein with honest accuracy. Certain names have been changed to protect the guilty ... and at least two innocent people.

As X Goes to Infinity
Copyright © 2019, 2025 by Amy Joy Hess
Created in the United States by Ocean to Mountain Publishing

No part of this book may be reproduced in any manner whatsoever without written permission except in the case of brief quotations credited to the author. For information, address Ocean to Mountain Publishing, P.O. Box 1116, Wallace, ID 83873.

Books published by Ocean to Mountain Publishing are available at special discounts for bulk purchases in the United States by corporations, institutions, and interested individuals.

Unless otherwise specified, scripture quotations are from the New King James Version, © 1982 by Thomas Nelson, Inc. All rights reserved. Used by permission.

Photographs and art are the work of the author unless otherwise specified.

ISBN: 978-1-962532-11-2

To my siblings.

I love you guys.

You are all amazing people,

and I'm honored to know you.

This book was primarily written during the years 2014 - 2015.

Table of Contents

	Cast of Real-Life Characters	1
	Introduction	3
Ch. 1:	The Albanian Ghost Child	7
Ch. 2:	A Little Levity	16
Ch. 3:	The Battle of Najaf	21
Ch. 4:	Splashing Life	28
Ch. 5:	The Iraqi Child	34
Ch. 6:	The Broken Drone	37
Ch. 7:	Stinky Bugs	43
Ch. 8:	Desks in the Bathroom	53
Ch. 9:	Plotting	65
Ch. 10:	Scrapes and Sprays	70
Ch. 11:	Turning Right	74
Ch. 12:	Spiritual Empathy	78
Ch. 13:	Deeper Dimensions	85
Ch. 14:	Plotting	91
Ch. 15:	The Zenith Party	95
Ch. 16:	Dimples	104
Ch. 17:	The Hollow Man	110
Ch. 18:	No Hammers!	114
Ch. 19:	Quantum Entanglement	118
Ch. 20:	Keep Smiling	125
Ch. 21:	Scroggin and Ferraris	131
Ch. 22:	The Grand Canyon	143
Ch. 23:	Fear	152
Ch. 24:	Angry Man at the Window	157
Ch. 25:	Zion	161
Ch. 26:	Sunshine and Prophecy	168
Ch. 27:	Stinky Bugs Reprised	177
Ch. 28:	Which God?	180
Ch. 29:	Life Presses On	189
Ch. 30:	The Life Inside Me	191
	Appendix	196

Cast of Real-Life Characters

Professors

Dr. Paul D. Stillwell (Geologist) - mentor
Dr. Zenith (Astrophysicist) - birthday boy
Dr. Jeff "Flash" Gurden (Biophysicist) - enduring friend
Dr. Bob Manchester (Biochemist) - excellent advisor
Dr. Vallo (Organic Chemist) - generous employer
Dr. Dan - Chair of the chemistry department

Brothers and Sisters

Lex - big brother, great guru in the fashion world
Heather - sister, who manages Starbucks and much more
Baron - brother (twin), tower climber, former military
Shadow - brother (twin), career military terrorist-stopper
Whitney - sister, gorgeous mother of a multitude, apothecary
Max - youngest brother, sports enthusiast and creative chef
Jordan - youngest sister, the horse-whisperer

Notable Schoolmates

Matthew Caerphilly - Physics partner, prayer partner
    (And fellow birthday planner)
Kate - Physics partner, "Apocalyptic Pale Horse of Death"
    (And fellow birthday planner.)
Stickley - Organic Chemistry pal with full, black hair
Michelle Caerphilly - lover of good coffee
Jared, Emily, Hannah, Kyra, Katie, Brennan, Don
    - adventure buddies on grandest Grand Canyon trip

INTRODUCTION

Fudge brownies. Rich, moist fudge brownies. Let's say we cut a pan of brownies into ten pieces. Each person gets 1/10 of that pan, and each piece is a decent chunk of brownie. Grab a tall glass of milk and enjoy - especially if you get an edge piece and it's chewy and even a little crunchy. Mmmm mmm mm.

On the other hand, let's cut that pan of brownies into 100,000,000 pieces. There's not much in one single 100,000,000th crumb of brownie. No glass of milk is needed there, just sadness and disappointment. As we get closer and closer to an infinite number of slices, we won't even get a proton of brownie.

$$\lim_{x \to \infty} \left(\frac{1}{x} \right) = 0$$

This looks like scary math, but it's not. To translate: the limit of **1/x** as **x** goes to infinity is **0**. What does that mean? As **x** gets bigger, the final result gets smaller. If we could cut our pan of brownies into an infinite number of pieces, our portion would be essentially nothing. That's the limit of **1/x** as **x** goes to infinity. It's zero. That's the end result.

When I first learned about limits back at age 17, I immediately associated **x** with Time. When it comes to Time and our universe, I envisioned the same thing. If **x** represents brownie slices, we get nothing. If **x** represents the disintegration caused by Time, then as **x** goes toward infinity, this whole world goes away.

The Second Law of Thermodynamics says that our universe is tending toward disorder. Entropy - randomness - always increases on a cosmic scale. The universe is ultimately breaking apart. As **x** goes to infinity, the Earth keeps spinning and spinning, flying around the Sun one millennium after the next. Billions of years pass, and the Sun swells into a red giant and finally explodes, destroying our solar system. All the stars in all the galaxies fly ever outward into the blackness. They eventually explode, collapse, burn out or otherwise die. Eventually, like our protons of brownie, every atom is scattered away from every other atom; the universe experiences a heat death as entropy fulfills its ultimate purpose and all energy is uniformly distributed throughout space. Everything we've ever known is gone. There is no life. There is nothing worth calling "anything." As **x** goes to infinity, this world is gone.

Let's say my atheist professors are wrong, and God lives. He is the answer for the miracles I've experienced and the First Cause of all that is. We don't have to wait billions of years. As **x** goes to infinity, He steps in and puts an end to this broken world. He's been stretching out the heavens,[1] and one day He's going to roll it all up like a piece of paper.

> *All the host of heaven shall be dissolved, And the heavens shall be rolled up like a scroll; All their host shall fall down As the leaf falls from the vine, And as fruit falling from a fig tree.*
>
> Isaiah 34:4

This whole universe is destined for destruction, no matter how we look at it.

And yet, let's say God rolls it all up, what's left? It's only the digital projection that's gone. The movie screen rolls upward, rising out of the way, and the eternal world suddenly comes into view. The new heavens and earth are presented to us, dimensions outside the ravages of Time, dimensions made to last forever.[2]

Secular scientists have a habit of looking at our visible space-time as the real world and the spiritual world as imaginary. As **x** goes to infinity, though, this world is zero no matter what. No matter what. This physical world is ephemeral. It's in the process of passing away into nothing. The only world that has a chance of being real is the spiritual world, the world that lies behind the curtain. That

---

1 Isaiah 40:22
2 Isaiah 65:17; Isaiah 66:22; 2 Peter 3:13; Revelation 21:1.

existence beyond space-time - that's the one that remains after this world is gone.

I'm betting on that real world. Mathematically, it's pretty useless to bet on this one.

CHAPTER 1

THE ALBANIAN GHOST CHILD

"Nevertheless do not rejoice in this, that the spirits are subject to you, but rather rejoice because your names are written in heaven."

Luke 10:20

So the nunnery was haunted. So what.

I walked past the doorway of the basement's dark boiler room, the monstrous metal beast silent in the shadows. I could imagine easing past that hulking furnace and into the dingy cellar room beyond. We'd been told that a tunnel once led from the cellar under the road to the old brick hospital across the street. True? Not true? The outline of an old opening filled with concrete remained in that back basement room.

Whatever its history, we took it for granted the building was haunted.

A week earlier, my 11-year-old brother Shadow trotted through the living room, where our handyman Ray sat swaying back and forth in the rocking chair. Ray had thick hair, prematurely white hair. Distinctive hair. Shadow continued past him down to the kitchen where Ray stood at the sink doing dishes.

"Wait," Shadow glanced behind him down the hall. "Weren't you sitting in the rocking chair?"

"What?" Ray looked up, his hands deep in warm water.

"I just saw you…" Shadow's eyebrows creased in bewilderment.

He turned and trotted back to the living room. The rocking chair sat empty.

Shadow wasn't alone. Five-year-old Maxwell played on the floor of Mom's bedroom when a stack of magazines caught fire on the table beside him. He knelt and stared at them in surprise. Mom walked in and saw the magazines aflame. "Ahh!" She threw a towel on the pile and put out the small conflagration. She lectured Max about the dangers of playing with fire.

Max always claimed he had nothing to do with it. Even as an adult, Max says the magazines spontaneously combusted. "I guess I did it," he still says, "but I do not remember setting those magazines on fire."

Before the nunnery, we'd been living in a camping trailer behind Gloria's Steakhouse in Prichard, Idaho. Six of us kids squished into that tiny trailer. Then, Mom's brother offered us a marvelous alternative; Uncle Doug said he wanted to relocate to North Idaho, and he'd buy a house if we picked it out. He said we could live there until he moved up.

Of course, Mom hunted up the biggest house in the county. She picked me up after school one day, excitement flashing Roman candles from her eyes. "I found it! I found our house. Guess how many bedrooms it has."

"How many?"

"Guess!"

"Five bedrooms?" I offered. I could tell by her enthusiasm it wasn't some little cottage.

She kept looking at me, waiting for me to try again.

"Six? ... Seven?" Had she found a castle?

"Eight!" she declared in triumph. "This place has eight bedrooms, 17 walk-in closets, five staircases, five bathrooms. It's… it's a crazy house. You have to see it. There's a bathtub in the wall of the master bedroom!"

"In the wall? What do you mean 'in the wall?'"

"It's in the wall. You walk up one step and you can climb into a working bathtub."

"A bathtub in the wall? Can we go there now?"

"Sure!" Mom said. "Let's go pick up the kids."

The old nunnery in nearby Wallace had been listed for $60,000, which was cheap in any decade. It had been built as the mayor's house half-a-century before the nuns moved in, and it was a grand old building. As we stepped through the front door behind the realtor, a wide staircase wound up around the corner to the second story. Hardwood floors stretched across the living room to the fireplace and into a little library by the broad front windows. We wandered past smooth, rounded corners and down a hall past the music room into a large kitchen with windows overlooking the back yard. Ten foot ceilings. We kids stared at the yards of cupboards stretching up and up and wondered who could reach that high to put dishes into them. The house just kept going on and on, with more hallways and bathrooms and closets and doors.

It was a weird old house. The coat closet in the foyer had a false back that opened into the first-floor bedroom. We hiked down another hall past the kitchen and up the narrow servant's stairs to explore the five bedrooms on the second floor. A bathtub really did sit in the master bedroom's north wall, just like my mother said.

There was more! A staircase in the second floor hall led to the third floor, but we found a hidden staircase in the east bedroom walk-in closet. We opened an extra door in that closet and followed stairs into the shadowy darkness to a trap door to the top of the house. A trap door! In our house! Shadow and I slapped each other's hands in delighted high-fives.

After the cramped space of our tiny camping trailer, we were ready for a 6,000 square-foot house. Sorry Uncle Doug that it needed some mending. Sorry that it turned out to be haunted.

We didn't care about ghosts. We danced all over the music room in the evenings, pop songs blaring. If we tumbled and wrestled onto the carpet, we didn't have to worry about smacking into end tables or chairs in the wonderfully empty room. We could run and slide down wide hallways in our sock feet and bounce into broad rounded corners without injury. And we didn't fear the massive basement with all its secrets.

Through the magic of YouTube, I recently watched a 1983 interview of Eddie Murphy on *The Tonight Show* with Johnny Carson. Eddie said something entertaining about ghosts:

> Eddie: There's the ghost in my house, that's why I'm moving.
> Johnny: A ghost?
> Eddie: Yes.
> Johnny: Does he talk or-?
> Eddie: No, see blacks aren't like whites with a ghost in the house.
> Johnny: Really? I didn't know that.
> Eddie: ...I was watching *Poltergeist* on telev- on cable. And I noticed that y'all just stay in the house when the house is haunted and -
> (Audience laughter)
> Eddie: "-we leave."
> Johnny: "You leave."

We didn't leave! A year after we'd moved in, I paused in the blackness of the basement hallway, thinking about the tunnel under the street. Then, I marched on into the laundry room to scoop warm clothes into the laundry basket, then clicked off the light and returned through the darkness. No evil spirit ever showed its face to me.

That didn't mean the spirits didn't bother other people. A single mother and her child lived with us for a couple of weeks the summer I turned 17. The young woman woke my mom several nights in a row, hissing in her ear, "Sheri! Sheri, I heard them again!"

Mom finally said, "Look Danielle. The ghosts aren't bothering me. You are."

Uncle Doug and Aunt Terri eventually moved up and overhauled the house, turning it into a bed and breakfast. Aunt Terri reports, "Every workman without exception who worked in this house would find me at some point to ask me if this house was haunted."

Something invisible walked loudly up the steps and down the halls, sometimes right in front of terrified observers. Something moved coats and hats and tools around. Something sat in the rocking chair, pretending to be a human. The bed and breakfast patrons came down every morning to eat, and each new couple asked (a little hesitant and embarrassed), "Um… is this house haunted?" Aunt Terri always answered, "Why do you ask?"

Of course, it isn't politically correct to believe in disembodied spirits. Dr. Stillwell frowned at Michelle Caerphilly and me when we started talking about ghosts one day before our Historical Geology class. The old geologist assumed that our ghosts were the stuff of wild imaginations. Perhaps the nunnery just creaked with age and settling. Perhaps people *expected* an old house like that to be haunted, and they only imagined the footsteps that clacked down the halls and stairs. Perhaps little Max set those magazines on fire. Perhaps.

Then again, we've had some noteworthy spiritual run-ins.

Some years after Shadow saw the fake Ray rocking in the rocking chair, his twin Baron was stationed in Vitina, Kosovo where his anti-armor platoon worked as a quick reactionary force (QRF) (82nd Airborne Division, D Company 1-325, Airborne Infantry Regiment), acting as peacekeepers between the warring Serbians and Albanians. One night, Baron walked into his tent to find his fellow platoon members gathered around a Ouija board. Specialist Sean Green had bought the board, and the guys huddled together, chatting with whatever spiritual things hung out there.

"Hey Truman!" one of the guys waved Baron over. "Check this out. We're talking to a 3-year-old boy and a 70-year-old man."

"Wow," Baron thought. "Those Albanian ghost toddlers can text, in English, via Ouija board. They sure make today's preschoolers look lame."

Baron declined and walked past the guys to lie down. He wasn't keen on talking to Uncle Screwtape and Wormwood junior. The guys continued speaking to the flimsy game board, and Baron grew increasingly anxious on his cot. Finally, too scared to stick around,

Baron left the tent. He wandered over to the small 24-hour chow hall, where he found the chaplain munching on an ice cream bar in the otherwise empty room.

"How are you doing, Truman?" the chaplain said.

"Uh. Not so good. I can't sleep because they're playing with a Ouija board in there."

"Well. Pray that God goes before you." The chaplain took another bite of his ice cream.

Baron nodded. Mom had said the same thing when she'd had to deal with demons. So, Baron paused and said a simple prayer. "Lord, go before me."

When Baron entered the tent this time, his fear evaporated. He returned to his cot to sleep, but he couldn't help but listen to his platoon talking to the Ouija board. He finally decided he'd have to get up and confront the thing.

"I shouldn't have talked to it," Baron says. "It was really stupid of me to talk to it. Once you're on their radar, it's hard to get them to go away. But, I wanted the guys to know that they were not talking to a 3-year-old boy and a 70-year-old man."

To use a Ouija board, people put their fingers on a little pointer, and it slides across the board to letters or numbers or to the words "Yes" and "No." Each person can suspect the jerk next to him is moving the pointer, so there's little certainty that one of the participants isn't faking the whole thing. Either way, it's dangerous to mess with spiritual doorways. Evil spirits are like Ebola. You keep yourself far far away from Ebola.

As I type these words, Baron is sketching trees on graph paper at my dining room table. "They kept asking it stupid questions," Baron keeps drawing. "Things like, 'What is it like to die?' and 'What's your favorite color?' Stupid things. Finally, I spoke up and said, 'Ask its name.'"

So, they did. They asked the board, "What's your name?"

It spelled out "D-E-V-I-L" three times in a row, and as they read off the letters, cries of alarm rose to the roof of the military tent.

I think that was interesting, that it showed some of its face like that. Demons hate to be honest about who they are. They're not extraterrestrial aliens. They're not the dead relatives we miss so much. They're not mothmen. They're not sweet 3-year-old Albanian ghost children. Demons are a bunch of posers. I also think it was lame of the spirit to go generic. C'mon. Give us your real name, Deception.

What took place that night was no party game. As Baron spoke to the entity in the Ouija board, the soldiers at the table couldn't hold their hands on the pointer. It shot across the board faster than they could keep up.

"It didn't even touch the board. It glided across the board like the puck in air hockey." He asked questions that an old Albanian man wouldn't know how to answer. "I asked it things the other guys wouldn't know. Things like, 'Who was the first ruler of Babylon?' and it told me, 'Nimrod.'"

At my table, Baron takes a drink of his coffee and goes back to drawing. "It did not like being questioned. It was obviously upset. It knew what I was ramping up to."

When it didn't want to answer a question, the pointer would spin all over the board in a blur. "I'd tell it, 'In Jesus' name, answer the question,' and it would unwillingly answer me. After awhile, it wouldn't answer any question unless I commanded it in Jesus' name. And then it would go back into its tizzy fit."

"How many of you are there?" Baron asked at one point.

"1."

"How many of you are there?"

"3."

"How many of you are there?"

"74."

"Why did you ask it three times?" I question him.

"I don't know," Baron shrugs. "I didn't believe it. I guess I was badgering the witness. But, I didn't ask it three times in a row - it was three times throughout the night."

That's another reason not to mess with Ouija boards. It's impossible to trust anything these entities say. They're skilled liars, skilled manipulators, and in Luke 4:33-35, Jesus told one to be quiet even when it spoke the truth. Don't engage unclean spirits.

Finally, Baron asked, "Who is God?" The spirit didn't want to answer. It spun madly around the board.

"In Jesus' name, who is God?"

"L-U-C-I-F-E-R."

"In Jesus' name, who is God?" Baron repeated.

The pointer flashed across the board in fury.

"D-I-E-D-I-E-D-I-E."

"I had to demand it to tell the truth in Jesus' name," Baron shakes his head. "Until it finally wrote out, slowly, like it really hated to say it, 'J … … … E … S … U … … … S.'"

I kind of wonder if there weren't angels there, forcing it at the tip of a sword like in a Frank Peretti novel.

Baron was getting tired, so he finally left to lie back down. He had wanted to sleep earlier that night when he'd first entered the tent, and now he'd wasted time on these evil beings that wanted him dead. He needed his rest.

"Later on when I went back to bed, they asked it some more questions," Baron tells me from the dining room table. "I could hear them. They narrated everything. They asked, 'Are you afraid of anybody in this room?'"

"Yes."

"Who are you afraid of?"

It pointed at Baron.

"Are you afraid of Truman?"

"Yes."

"Why?"

"G-O-D-G-O-D-G-O-D."

Baron didn't perform any holy rites. He didn't join the priesthood. All Baron did was ask God to go before him. That's interesting. I also think it's interesting the entity so often gave its answers in triplicate. I wonder why that is.

Finally, Baron had to get back up and order the spirit(s) to leave. He rests his pencil on the table and looks at me. "I said, 'I bind you in the name of Jesus, and I cast you out of this board. You are not allowed to come back ever again. None of your friends are allowed to use it. Nobody else down the line is allowed to use it.'"

The night terrified the other guys – tough military guys. Greg Allen asked to borrow Baron's Bible so he could sleep with it. "Nobody made fun of him for it either. They all acted like, 'Gee. I wish I'd thought of that first.'" Baron kindly handed it to Allen, even though it's not the paper and ink of the Bible that saves us. It's the truth it holds that matters. But, anyway, it made Allen feel better.

People are nuts, though. The fear didn't stop the soldiers from wanting to try the board again.

"I wanted to buy the Ouija board from Sean Green," Baron says. "I offered him $60 for it, because I wanted to burn it. Green didn't want to sell it to me, though, because he'd just spent the night watching an evil spirit talk to people, but it never worked again. Months later, Green came up to me and said, 'Hey, Truman, do you want to buy the Ouija board?'

"I told him, 'I'm not buying that from you. It's just a piece of cardboard.'

"'Well, what should I do with it?' Green asked.

"I said, 'Burn it.' So, he did."

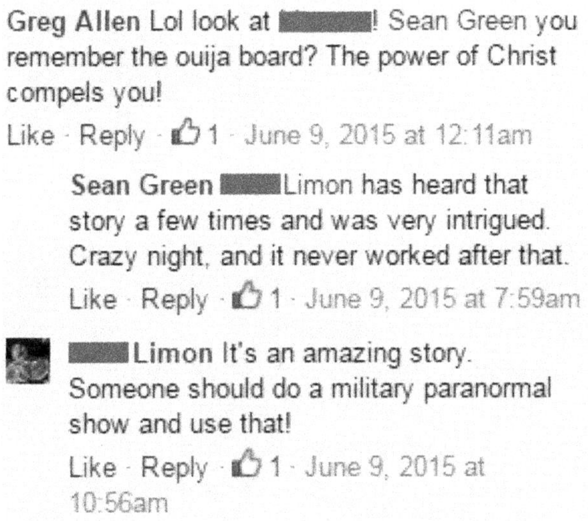

> Greg Allen Lol look at ▮▮▮▮! Sean Green you remember the ouija board? The power of Christ compels you!
> Like · Reply · 👍 1 · June 9, 2015 at 12:11am
>
> > Sean Green ▮▮▮Limon has heard that story a few times and was very intrigued. Crazy night, and it never worked after that.
> > Like · Reply · 👍 1 · June 9, 2015 at 7:59am
> >
> > > ▮▮▮Limon It's an amazing story. Someone should do a military paranormal show and use that!
> > > Like · Reply · 👍 1 · June 9, 2015 at 10:56am

Figure 1: Found on the company Facebook page in 2015.

Chapter 2
A Little Levity

Dr. Gurden: You rub a balloon on your head and stick it to the wall – why does that happen?
Students: Magic.
Dr. Gurden: Yeah, that's what they probably thought when balloons and walls were invented...and hair.

The best things of life don't always look exciting at first. They happen while normal life is going on. They *are* normal life. Spring semester of 2011 started off like any other. I didn't suspect the fun that lay ahead those next few months, and I had no idea how greatly the spiritual realm would invade my safe little science department.

I sat at a black lab table before Dr. Gurden's Physics II class, safe inside on a cold January day. I had reached Flash's 9:00 class early and carefully stacked my pencils into a small pile. A lone Lego man eyed me from behind the pencil barricade, a gravy-stained wife beater under his suspenders. He wore a plastic-leather bike helmet with the goggles propped up on his head, and stubble mottled his chin. Gone were the days of simple yellow faces. The bin of Legos we'd found under our Christmas tree included tough Lego guys with helmets and beards and sunglasses. There were Lego zombies with red eyes and knights and military dudes ready to get

Figure 2: Gravy-Stain Lego, hefting a rocket launcher.

their rage on - ready to rage on physics with Dr. Gurden.

Physics was my *fun* class that Spring 2011 semester. I got to take it with Jeff "Flash" Gurden, whom I regard as one of the world's heroes. He really is. He's a hero. Times on geology trips when students might have mutinied against Dr. Stillwell over his tendency to change plans last minute - or fail to reserve camping spots in time - or get flight dates wrong... Dr. Gurden's intuitive wisdom cooled the glowing violence of frustration within the student ranks. Even this early in our relationship, I had hiked and camped and played cards with the school's young physics professor. One day, he'd instruct me in the art of roasting the perfect marshmallow, and we'd have long discussions in the rain over the mechanisms of life. Even then, in January 2011, he brightened my day.

This spring semester was set up to be the left-brainers paradise. My other classes included Physical Chemistry, Organic Chemistry, Instrumental Chemistry, and Statistics. Seriously, Physics II was the most creative and colorful class in that whole bunch. Not that I'm knocking Instrumental Chemistry. I mean, I make my living on that stuff now, and chemistry professor Dr. Dan was a good teacher. But, this was one of the few semesters I didn't take screenwriting or radio practicum or the "Scottish play." Another 11 months would pass before I stumbled onto stage as Shakespeare's drunk porter in *Macbeth*. "Knock knock! Who's there!" I desperately needed some levity.

Flash spotted my Lego biker. "Hmmm, Amy Joy. Aren't you a little old for Legos?"

"What! You're never too old for Legos. Do you want one?" I reached into the right front pocket of my jeans. "You look like you need a Lego man, Dr. Gurden. Here you go, you need a ghost."

"Really?" Flash raised his eyebrows. "You mean, I can have it? To keep?"

"Sure! We have tons of them. See? Its little plastic ghost sheet comes off, and it's got the guy underneath?"

"That's really neat, Amy Joy," Flash said. "Thanks."

"But you have to keep it," I told him. "You have to enjoy it.

Don't hide it away somewhere."

The summer before he died, Randy sneaked those Legos home in a black garbage bag. He'd scored pounds upon pounds of the foot-wounding blocks from a young adult male in Pennsylvania, and he'd hidden them away with plans to wash them up and give them to our three kids. Alas, the best laid schemes o' mice an' men, gang aft agley, if you know what I mean.

I'd finally hauled the Legos out of storage in December and washed the piles in a washer at the laundromat. (Tie them up in a pillow case or two and swoosh swoosh swoosh.) Dumped them into an empty Shop Vac tub and rolled them under the tree on Christmas Eve with a bow. We were in Lego heaven in that house, baby. A late Merry Christmas from Dad.

Flash did *not* hide his new Lego ghost; he combined it with a bicycle wheel to teach centripetal force! I'm honored and pleased.

After New Year's Day, Dr. Vallo's wife, Kathy, had called on me to clean her house again and help rid it of stink bugs. I vacuumed every room, sucking up the little brown invaders under beds and behind chairs. Dr. Vallo also sent me to his mother's farmhouse to clear ivy away from her old chicken coop.

When Dr. Vallo's ancient Hungarian parent invited me in for tea, she grumbled, "I want to eat white English muffins. I do not like ze whole wheat! And I like my salt! I am 90-years-old, and I could die any day, and I want to enjoy what I eat. Kathy makes me food, but she does not use ze salt. She is so afraid of it!!" I smiled at Ildikó's food frustrations as we enjoyed our tea together.

Yes, the children and I had a pleasant, productive break.

Legos are important. Stink bugs are important. Ghosts are important. I'm not telling you these things for nothing.

On January 4th, before the semester started, Dr. Stillwell invited me into the school to cut tiles for my kitchen floor. When I'd arrived at his office door, I found Dr. S. on the phone with somebody about his computer, so I remained politely outside. As I waited, his conversation floated into the hall, and I giggled at the turn it was taking. Dr. Stillwell must have heard my chuckles, because he grew cheerier.

"Well, Amy Joy," he finally turned to me. "How did your Christmas go?" We sat calmly and chatted about our holiday experiences. It was warm and pleasant, and we enjoyed catching up after the month-long break.

After a bit, Dr. Stillwell rose and patted my arm and walked past me into the hall. He always acted like he simply provided stability to my chaotic life, but I smiled as I followed him down the hall to his lab. He had missed me. I knew it.

We filled his rock saw reservoir with water, and he got me all set up to do great things for my kitchen floor, but then he lingered as I prepared to buzz through the tiles.

I puzzled at him. "Don't you have to go to lunch?"

He shrugged. "Oh, I'm sure I have chores to do…"

His unwillingness to disappear pleased me. He made his exit with reluctance, and I made good use of the rock saw, but I grinned inside. I liked hanging out with the old geologist, and it was nice that he enjoyed my company and left only because he had to.

As soon as I finished my work, I took the opportunity to cover his Power Point station with Lego military men. Haha! Within a few minutes, a small Lego army had invaded the geologist's lab computer, complete with with swords and battleaxes, and I secretly hoped they'd hold their ground there all semester.

"You can't keep them, though," I told the good doctor when he returned and discovered he'd been invaded. "You don't accept gifts from students!"

That was early January. Now school had officially started for the semester, but as I waited for Flash's class to begin, I reached into my pocket for another Lego man and finished my pencil barricade.

Right about then, the geologist himself wandered in and made his way to my spot near the white board. Dr. Stillwell gazed down on my little Lego thug behind its barricade and shook his head. "I'm not surprised," he said. "You know, it's usually senile old people who regress to childhood."

Lego guys soothed my soul, and I didn't require his approval. Besides, I could tell that in his heart of hearts Dr. Stillwell wanted a Lego guy. I could tell because he got all kinds of jealous when he saw Flash's ghost.

"What! She gave you one!" he barked. "She didn't give me one! She's a Native American giver!"

The real point is *not* that I was a full-grown adult playing with Lego knights and brutes before my physics class. The point is that I enjoyed lighthearted fun with my professors.

I *require* silly fun! My mind can't decide whether it likes math and logic best or whether it prefers I write goofy poetry and act on stage as the drunk porter (which, of course, is the preeminent role in *Macbeth*). I enjoyed Physical Chemistry and all that cool stuff about thermodynamics, and I enjoyed Instrumental Chemistry and all that cool stuff about atomic absorption and mass spectroscopy. I looked forward to Physics II with Flash. But, I was going to explode and kill somebody dead if I didn't have some creative outlet.

Biochemistry professor Dr. Manchester always said, "kill you dead" in his wonderful British accent. It was his favorite saying. He'd explain, "So, if the hemoglobin held onto the oxygen and didn't release it to the myoglobin in your cells, it would kill you dead." Killing dead was going to happen if I didn't find something creative to do that incredibly left-brained semester.

This book is not about Legos and cabin fever, and it's not *just* about my adventures with my charming professors. It's about parties and practical jokes. It's about conquering fear. More than anything else, this book is about the tendency of the spiritual world to invade the four dimensions of our visible, physical space-time domain. I was a science student surrounded by chemists and geologists, and God's guidance and power burst into my relationships with these men that fun-loving spring.

CHAPTER 3

THE BATTLE OF NAJAF

His friend responded, "This can be nothing other than the sword of Gideon son of Joash, the Israelite. God has given the Midianites and the whole camp into his hands."

Judges 7:14 (KJV)

On January 28, 2007, the mayor of Najaf was attacked as part of a power play by terrorist forces. This was back when the United States still had a strong presence in Iraq. The climax of the Shia Muslim holiday of Ashura was approaching, and a group of terrorists intended to take over the city and declare their leader the 12th Imam. A small group of Iraqi soldiers and their Green Beret U.S. advisors had been getting ready to leave the city that day, but the attack delayed their plans.

It's easy to find grief or suffering anywhere we look in this world, because we live on battlefield Earth. In the midst of the traumas, though, God's protection and provision persist. Sometimes His demonstrations of power are huge and obvious, and that January of 2007 my brother Shadow lived through a miracle of international significance that every bit echoed Gideon's victory in Judges 7.

Sergeant First Class Donny told me, "We were just about to head out when we got a call asking if we could help the mayor get out of this little firefight he was in."

The U.S.-led team spent an intense three or four hours fighting it out with the Soldiers of Heaven, a Muslim cult, but the U.S. mission was successful. They rescued the mayor and transported him to safety, then returned to their Najaf operating base, "and got plussed up with ammo and food and gas."

Shadow told me, "We were thinking, 'That was bad.' Because a lot of people got shot. Sean Kirkwood got a Silver Star for that morning, and you have to practically save the president's daughter to get any kind of recognition in Special Forces." The team didn't know that the "bad" of the day hadn't even started.

The Green Berets don't usually advertise their work, and it's been difficult to find comprehensive information about that day's events. The Melbourne, Australia newspaper *The Age* published a decent article,[1] but *The New York Times'* version sounds like official propaganda from the Iraqi government, written to sound more in control than it was.[2] I don't know…maybe *The New York Times* was right and there was a midnight meeting and a battle plan to take out the Soldiers of Heaven Muslim cult, but I have serious doubts based on accounts of the different battles from soldiers who were there that day.

There are many things the media got mostly right. There were about 250 terrorists killed and several hundred captured during the battle that took place that afternoon. The Soldiers of Heaven militants in Najaf believed their leader to be the 12th Imam, the Muslim messiah. They were planning to take over Najaf during the Ashura holiday. Two Americans died in a helicopter crash. Eventually, U.S. Strykers and air support arrived to help, but not until many hours after the fighting had started and the battle was basically won.

I cannot find a single article that explains that the battle was fought by about 62 U.S.-aligned forces. The American side was outnumbered at least (at *least*) 10-to-1, yet they won an absolute victory.

This is a big deal.

Donny and Shadow and their guys included a group of 20 U.S. Special Forces and 30 Iraqis they had trained, plus another ODA of 12 or so guys who showed up there from Hillah. This little Gideon's

---

1 US, Iraqi forces kill 250 militants in Najaf. (2007, January 29). *The Age*. Retrieved May 6, 2015, from http://www.theage.com.au/news/world/us-iraqi-forces-kill-250-militants-in-najaf/2007/01/29/1169919241141.html.

2 Cave, D. (2007, January 29). Iraq: 250 Insurgents Killed in Battle. *The New York Times*, retrieved May 6, 2015, from http://www.nytimes.com/2007/01/29/world/middleeast/29iraq.html?_r=0.

army was ambushed by a group of 600-1000 enemy soldiers after the helicopter crashed, and nearly everybody in that little band of Iraqis and Americans came out alive. Even the newspapers agreed that the other side took all the losses. Even the newspapers got that right.

The U.S. forces were scheduled to leave Najaf that 28th day of January. They were getting out, which tells me that nobody was planning to attack the Soldiers of Heaven, no matter what *The New York Times* says about midnight meetings. The first time the Green Berets tried to leave, they got called on to rescue the mayor. The second time, they got a call: "Hey, we got an Army bird down." The Apache helicopter was shot down first, and the U.S. team was sent out to recover the bodies of the men inside.

They set off toward the smoking helicopter. "It was kind of weird, with fire coming from all over the place, people taking shots from alleys or berms. The desert's kind of crazy," Donny told me.

"There's a funny story," Shadow said. "There was a small gun fight on the way. We were driving along, and one of our Iraqis was shot off the back of the Humvee, and we almost ran over him, so we stopped and got him. And while we were stopped, I ran over to his Humvee and started working on another Iraqi who had been shot in the hip. I shoved him up into the back of the Humvee, kinda like the back of a pickup truck, and I'm standing there working on his hip to stop the bleeding. All the Iraqis are huddled there, staring at me, and I'm shouting to my Iraqi medic, 'Hey! Put your hand here! Put pressure right here!' And they're all staring at me. Later, I learned there were bullets flying all over the place past me, and I'm standing there working on this guy, completely clueless."

"They thought you were purposely protecting them."

"Well, I was trying to patch him up! He said later, 'Shadow! You are like my brother! You must love me so much, to take care of me and put me behind cover while you were out there exposed! You are so brave!' They all thought I was really brave. As far as I was concerned, I was just patching the guy up. They didn't realize I came from a large family and can tune out noise."

The team moved on and decided they'd drive around a big dirt bank and out to the helicopter crash site. "We had to drive between two berms," Donny said, "That's not good; it's a perfect place for an ambush."

There didn't appear to be another route to the helicopter, so the trucks carrying the few U.S. and Iraqi soldiers drove into the danger zone. As soon as they did, the Soldiers of Heaven militants poured over the tops of the berms and began shooting at them from yards away.

Shadow said, "We picked a route, and it just happened that route went through the center of the Heaven's Army. And right there is the 12th Imam surrounded by all his people."

"They used the berm as cover and started shooting at us, basically at point blank range," Donny said. "I was inside the Humvee behind bullet-proof glass, but the guys like Shadow in the back of the truck, they were only protected on the sides. The back was open and totally exposed."

I never worried about my brother Shadow, no matter how many times he was deployed to the desert. Mom had lost her first husband in Vietnam, and she hated that Shadow was repeatedly sent into the warzone of the Middle East. She prayed for him all the time, but I never worried about him. The U.S. forces were good. They hunted down terrorist leaders, broke down their doors in the middle of the night, and arrested them - sometimes zipping ties on their wrists while the groggy militants were half asleep. Baron and Shadow had both done tours in Iraq, and they'd come back just fine.

Then, in late January of 2007, it occurred to me that Shadow could die. I had just stepped over the threshold of my cabin. I had something in my arms. It might have been baby Zeke, or it might have been groceries. I was in the middle of living my life when I was hit by an overwhelming sense that Shadow could actually get killed.

So, I stopped.

I must have closed the door behind me to keep out the cold, but I stood in my entryway and prayed before walking another step. I asked God to protect Shadow and all the guys with him, to keep them safe. I prayed for maybe a minute. Then the burden lifted, and

The Battle of Najaf

I went about my day.

I later learned that Baron and Dad had similar experiences at that time; they suddenly felt a strong urge to pray for Shadow. It makes me wonder how many other parents and friends and family were called on by Heaven to stand in the gap for those 60 young men ready to be slaughtered by a cult leader who thought he was the 12th Imam. I think of the people out there praying for our soldiers every day.

"We were fish in a barrel," Donny said. "They had a little mini-city they were building there, and we drove right into the middle of that hornet's nest."

Not only were the Iraqi and U.S. soldiers hugely outnumbered, but the Soldiers of Heaven were hopped up on drugs that made them feel *invincible*. It's hard to kill people whose bodies refuse to believe they're dead. Little prayer packets were found with the bodies, which they had thought made them *invisible*. No terrified, running Arabs here.

"These guys were high," Donny said. "I'm looking directly into these guys' eyes through my bullet-proof window, and they were high."

Shadow agreed. "They weren't just high. They thought we couldn't see them. They were brazen!"

Air support did arrive eventually, but it took hours because of radio frequency issues. In fact, the battle was being watched from the air, but the messages to send help weren't getting through.

Shadow said, "It was so bad - what they were watching on the overhead view from a drone - a supply guy was told to get a bunch of cots and body bags, because they were expecting us all to be dead."

"But you guys didn't get shot?"

"It's so weird, I didn't get a scratch," Shadow told me. "Not a single scratch. And this is like, you put your finger in the air, and it gets shot off. It was unbelievable fury! Unbelievable fury. The Imam's guys were so close we could reach out and touch them. I was definitely red most of the time on my ammo."

"Can you believe that?" Donny said. "That's God. Generally, if you set up an ambush, you should kill somebody. I'm not saying they were bad, because it was a pretty good ambush, but we had God."

The enemy used everything. They had .50-caliber machine guns, which are devastating. Audie Murphy famously won the Medal of Honor in WWII after he held off six tanks and 250 German soldiers for more than hour with a .50-cal machine gun. The Soldiers of Heaven also had 25mm quad Gatling guns and anti-aircraft artillery. Shadow estimated that 25 rocket propelled grenades (RPGs) hit his truck without exploding.

That's unfathomable.

Shadow said his team didn't do the assessment after the battle, but they figured that several hundred enemy soldiers were killed and several hundred more were arrested. In contrast, the U.S. side lost two Iraqis during the fight, two Iraqis from the group they were advising. Shadow grieved over these two men he had trained, men who'd become his friends.

The U.S. team got past the berms and to a more defensible location, but the serious part of the battle lasted from about 1:00 p.m. until nightfall. After the fight was over, Shadow and Donny got down on their knees and honored the Lord.

The Army later learned that the Soldiers of Heaven had been intending to attack Najaf the next day as thousands of people marched in honor of the Ashura holiday. The cult leaders had 100 women and children from their own camp ready to act as suicide bombers during the Ashura march, which would have resulted in mass carnage and devastation. The cult was a Shia cult, and Ashura is a Shia holiday, but the plan was to blame the Sunnis. The mass slaughter of Shiites on Ashura would have destroyed political relations between Shiite and Sunni leaders and pushed the country into all-out civil war. The destruction planned for that January day was halted by the presence of the U.S. forces who trusted in a different God.

Dr. Stillwell didn't want to hear this story when I tried to tell it. He sneered, "Oh. The Americans won because God was on their side."

Hey. In the name of his god, the self-declared 12th Imam

was willing to have 100 suicide bombers destroy their fellow Shia Muslims, but because Shadow and Donny's team rolled into that ambush, 100 women and children were spared death as suicide bombers, and their thousands of planned victims were able to walk home.

As one of the terrorist soldiers lay dying, he insisted that the Army of Heaven had won.

A U.S. soldier standing by him said, "What are you talking about? Your people all died."

"It doesn't matter whose blood is spilled," the cult member said, "just as long as blood is spilled." That tells us something about the god served by the Soldiers of Heaven.

After the battle, Shadow's shirts were so trashed with bullet holes and enemy blood, he threw them away.

I said, "What! You should have kept them!"

He said, "They had holes all through them. At the time, I was just thinking, 'Oh. These are useless. Somebody will give me crap for being out of regs for wearing them.'"

On the other side of things, Shadow was humbled by the fact that the Holy Spirit had urged people to pray for him. "It meant that I could have died. I didn't have to survive. But, because I did survive, I get to be grateful that I was allowed to. God's plan doesn't require me, but He wants to involve me, and I'm grateful for that."

He finished his story with a song lyric. "Even though my eyes can't see, Lord thank you for including me."

Shadow was awarded a Bronze Star with a "V" device for valor after that day in Najaf. Man, I wished he'd kept those shirts with the holes in them.

Chapter 4
Splashing Life

Those in power were beefing up the physics prowess of the university. That is, they were hiring an additional physics professor, and that meant we were asked to watch professor tryouts. Professors had to audition for the job, and it was a good thing because there are brilliant people out there with no communication skills. That's a painful experience – sitting in the classroom of a genius who can't teach worth dead stink bugs.

So, the department held professor tryouts! The candidates each presented us with an example physics lesson, and we students attended to provide them with a live audience. I did my duty and attended several of these lectures and even brought my children. Four-year-old Zeke sat on a stool at a black lab table in Dr. Gurden's lab and contentedly watched scholars talk about optics or cannon trajectories. Zeke was such a cool kid.

The series of potential professors all disappointed me, though. There was something missing in their lessons. I kept looking for a quality I wasn't finding; every single candidate lacked a certain "something." I didn't even realize I was hunting after it; I just knew that Dr. Stillwell seemed pleased following one of the sessions, while I had been bored in my soul.

"Wasn't that one good?" he asked me.

I shrugged. "It was okay."

It *was* a good session. The potential professor had done a great job of engaging people, of keeping my little kids interested. The lesson was presented in an easy-to-grasp fashion, and the information was stimulating. I just felt bored in my soul during it.

That struck me. It struck me that Dr. Stillwell didn't feel bored in his soul. And I did.

What was I looking for? What had I been seeking and not finding? I pondered over it, thinking about it, trying to figure out exactly what each of those potential professors lacked. I finally concluded that I was hunting for candidates who had *life* in them, who had so much life inside of them that it splashed on the rest of us. I didn't find that in any of the candidates, which really smacked me in the nose. Because somewhere along the line I'd known people who had splashed me with life, and these physicists didn't do it.

As a matter of fact, Dr. Stillwell didn't splash me with life either.

I loved Dr. Stillwell, with his wit and merry jokes and kindness toward me. We were planning a school trip to the Grand Canyon for that summer, and we muddled together over specific dates and routes. He overviewed the national parks we'd visit. We joked and argued, and he got constantly frustrated when I didn't immediately trust him about decisions he'd made. I appreciated him dearly, and he filled a horrible emotional hole I'd had ever since my brother Lex left home. Still, when I sat in Dr. Stillwell's presence for too long, it was like he sucked the life out of *me*. I had to recover from being around him. What was that about?

I began to think about it, to pay attention to it. I threw it out to my fellow Physical Chemistry students one day.

"Do you guys find there are some people who fill you up, who make you feel more alive? And then there are other people who drain you?"

Stacie understood. "Yeah, I know what you mean."

"I must be a drainer," Stickley announced. "Because I'm always pumped up by everybody. I'm draining their life force. That's why my hair stays this nice color."

We all laughed. Stick did have a full head of black hair.

I started paying closer attention to what made me feel alive and what made me feel dull and dead and empty. It wasn't something new; I'd experienced these sensations most of my life, but I'd never really tried to nail them down before.

I readily admit that I'm an introvert, but nobody ever believes me. They all say, "You're the most extroverted person ever!" Understand that I'm an extrovert by training, by practice, by forcing myself to speak up. I can chat with strangers and speak on stage and tell stories to crowds, and it's easy because I've worked hard at it. The reality is I'd just as soon go hide in my room and curl up on my bed and read. Being alone is safe and calm and peaceful.

There's more to this than an introverted personality, though. It's not just a matter of enjoying the company of people and being charged up by them. The very presence of certain individuals gives me a sense of life, and the presence of others give me a sense of deadness. I can walk into the room and feel it. I can feel it when I talk to them over the phone.

The deadness is like… it's like a heaviness or a hollowness, like I have pneumonia or pollen clogging my lungs. That's as close as I can come to describing it. It's unpleasant. I don't like it. The alternative is a feeling of life, of deep contentment. It's comforting. It's warm. It's filled with health and wholeness. It's fulfilling and satisfying. And sometimes it gets all fiery in my chest.

Dr. Stillwell drained me. It didn't matter how dear he was to me, how much I loved him, how often I would invite him over for coffee if I could. It didn't matter if we were inside or outside, in the sun or the rain, in a little office or a big cafeteria. It didn't matter the location - with him or with the others who made me feel that way.

I don't ever tell people, "Oh honey, did you know you're empty inside? Do you know you're a shell?" Sometimes Dr. Stillwell has felt *so* hollow, I don't know how he can stand it, and I'm certain everybody else in the whole world can tell. I feel like they all know he's an empty peanut, and I'm embarrassed for him.

I'm a scientist at heart, and I live and work in a world of scientists. I am fully aware that these feelings of mine are *subjective*. I can't put them in a culture tube and run tests on them. I can't tell anybody else how to duplicate them. They are my own subjective feelings, however. Which means that I can make constant observations about them for myself. I can mark down the specific circumstances that seem to cause them and describe them as well as I can.

If my appendix were inflamed and I felt pain because of it, that too would be a subjective feeling, but I could describe it and the specific circumstances that caused me to feel it. Other people have appendixes too. Other people can share their subjective feelings about their inflamed appendixes, and we can compare notes.

I have a spirit, and other people have spirits as well, and I'm thoroughly certain there are people reading these words who know exactly what I'm talking about. They've felt the emptiness in other people, and they've felt the life. Maybe we all can sense it to some extent. Maybe we should all pay closer attention.

I started noticing these things, things that I'd felt my whole life but hadn't bothered to study.

Then came a day in Physical Chemistry when I knew I was supposed to pray for Stickley. Stick, with all his nice black hair. I sat watching Dr. Li pen formulas up on the white board, and with no warning at all, I was overwhelmed by an urge to pray for Stick sitting in front of me.

I said to the Holy Spirit, "I can't pray for him right now. That's weird."

At that, the feeling of deadness, of emptiness, filled my chest.

I thought about it… about reaching out and setting my hand on the silly young man in front of me. He was wearing a black cotton t-shirt, and it would be so easy to rest my hand on his back and pray for him. As soon as I imagined it, my chest filled with a burning - a lively, healthy burning.

"No," I said again to that burning in my chest. "We're in the middle of class! I'll wait until after class."

Deadness.

"Okay…maybe I can pray for him now."

Burning.

"That's so ridiculous! I'll wait until later."

Deadness.

You'd have thought I could have just *said* a prayer for Stickley, but, that clearly wasn't enough. I had to put my hand on his back and pray for him right then. Right there in class. There was no guilt involved, as though I had some niggling sense I was responsible for Stick's life. I didn't feel responsible for Stick's life. I just knew that I needed to pray for him right then while Dr. Li stood up front marking out a long list of calculations. That's what I was supposed to do, and whenever I considered placing my hand on Stick's back, the burning feeling filled my chest.

But I didn't do it. I chickened out.

I know! I'm upset at myself! Years later, I still wish I could go back and do what the Holy Spirit told me to do!

Instead, I played it safe and decided to catch Stick after class and pray for him. That's what I planned, but of course I didn't get the chance. As soon as class ended, he hopped up and disappeared into the crowd. I tried to follow him, but he got away, and I didn't see him again for a week. An entire week.

He'd been crushed by a tractor.

Just kidding.

But arghhh! I'd screwed up! There was something important going on, something that required hands-on prayer, and I didn't do it! How much was at stake? Was it something fixable? I hoped so. I didn't want my cowardice to have ruined an important moment in Stickley's life. But, maybe it *did*, and it grieves me to this day.

Stick returned seven days later, and I sat behind him again in Physical Chemistry like usual. This time I reached out and placed my hand on his back, and he didn't even turn around to look at me when I did it. I prayed for him even though I knew I'd missed the significant moment. How many times has God given me Isaiah 51:12!

> *"I, even I, am He who comforts you. Who are you that you should be afraid Of a man who will die, And of the son of a man who will be made like grass?*

Over and over. He's given me that verse so many times! Why do I continue to worry what people think if God's told me to do something!

I approached Stick with it during our next Organic Chemistry lab.

"Remember when I put my hand on your back in class last week?"
He nodded.

"Well, I was supposed to pray for you in class the week before." I told him about the alternating burning and the deadness. "I was supposed to pray for you right then, but I didn't. I thought I'd catch you after class, but you left so fast - and then you were gone for a week. Anyway, I just wanted to apologize."

"Two weeks ago?" he said.

I nodded.

"Hmmm. That's when I was gone for training." Stick was in the military. "It was a horrible week. It's all your fault, huh?"

I nodded, smiling. We both smiled. At least he was still alive, but I have no clue what spiritual battle was going on. I have no idea how I failed Stick or the additional chain of people the prayer could have affected. Does God have contingency plans? I'm sure, but the Holy Spirit was insistent, and I didn't take Him seriously enough.

It wasn't the first or only time I've been a coward. I don't know what God's doing in people's lives, but it doesn't matter, because the specifics are none of my business. I want to do whatever the Spirit of God asks of me, because He's always brilliant and He loves people! I know I can't trust myself, and so I pray that in the future I'll not fail. God will forgive me, but that's not the point. He doesn't direct me if it's not important.

The burning and the deadness wasn't exactly new, but it was a new form of guidance. After I chickened out on poor Stick, I paid even closer attention.

Chapter 5

The Iraqi Child

My brother Baron had a much worse experience when he was a check point guard in Iraq in 2004. He'd finished up his time with the 82nd Airborne and had joined the National Guard to help with school funds. Almost immediately they sent him to Baghdad.

It was a miserable, blistering Iraqi day. The normal long line of people waited to file into the safety of the Green Zone, and Baron manned checkpoint #3. A multitude had spent hours in the hot sun before they reached him, but he had learned enough Arabic to tell people what he needed them to do.

"It's amazing how easy it is to direct people when you speak their language," he told me. "Some guys would shout at the people, which I hated. We were there to *help* the Iraqis. You know what? Other checkpoints got bombed while I was in Iraq, but not my checkpoint." Baron purposely treated the Iraqi people with respect, and they appreciated it.

"Please stand behind the line. Please put out your arms. Thank you. It's hot today, would you like some water? Please take the battery out of your cell phone. Thank you. Everything is fine. Have a good day."

Up through his line, there came a woman holding a young child. The little girl had been severely burned in an explosion somewhere, and a hand and foot had been burned off. Her face was charred, and she'd lost an eye. The woman had been waiting all day in the boiling heat, waiting to get to doctors in the Green Zone. As she came near, a crowd of angry people began shouting farther down the line, which meant that the gate had to be shut. By regulation, the gate had to immediately close at the warnings of a riot.

Immediately Baron approached the woman at the fence, stretching through the wire with a bottle of water, trying to get her and her little one the precious life source.

Just as Baron reached with the water, a reporter marched to the front of the line and shoved the woman with the burned child to the side. "I have a meeting. I have a meeting with General Bremer." Was it a BBC reporter? An CNN reporter? It didn't really matter. Rage flushed up in Baron.

"You have to go to the back of the line like everybody else," Baron told her.

"But I have a meeting with General Bremer," she snapped.

"General Bremer would tell you that you have to go to the back of the line."

"I have to be there in 10 minutes."

"I don't care!"

Baron told the woman with the burned child, "Five minutes. Wait five minutes and we can let you in."

As those five minutes passed, Baron felt the strongest urge to pray for the little girl. He had to lay his hands on that child, and it was urgent. "You need to put your hands on her when she comes through the gate," the Holy Spirit told him. "You need to put your hands on her and pray for her."

Baron argued, though. He was afraid.

Here he was, a seasoned soldier who had already served in Kosovo. Every day, he checked people potentially loaded with explosive devices. He daily faced the threat of being shot, and he'd had mortars and rockets just miss him. Yet, God asked him to pray for a little burned girl, and he worried about offending the people. He was just afraid.

"Do it," the voice inside him said. "Pray for her. You won't have much time. Put your hands on her and pray for her."

But, he didn't. He didn't lay his hands on her. He got the mother and child water. As soon as they came through the gate, he called the medics to attend to her. The girl received medical attention, but Baron was supposed to pray for that burned child, and he didn't.

It's one of his biggest regrets about his time in Iraq.

There are things in life we wish we could go back and do *right*. There are things we wish we could fix. Baron has no clue what happened to that little girl, but she was important. She was important enough for him to set himself aside and take 20 seconds to bless her. She had that value, and he failed her.

We all have to care about each other more than we care about looking silly. We have to notch it out in our heads that nothing is as important as simply obeying God when the moment comes, no matter what. Because when God asks us to do things, it's because they really matter.

Chapter 6
The Broken Drone

"There's a part of me that suffers from Tourette's of honesty."

-Dr. Stillwell

Dr. Stillwell had wanted to study bryozoans with me, and he'd suggested I do my senior research project on the Mississippian bryozoans he had stashed in his back lab room.[1] He approached me with it over Christmas break – it was his idea – but the next week he apologized and told me he didn't have time to help me. It was this demotivational whiplash. "Hey! I have an idea! You should do your senior research project on bryozoans! Wait. I have no time. I'm really sorry."

We had to wait until the summer to attack the little zooids, but that didn't mean we couldn't make other fun plans that semester. Like a class trip to the Southwest.

The Grand Canyon! A trip to Grand Canyon also meant side adventures to exceptionally fantastic places like Bryce Canyon and Zion National Park in southern Utah. I adore Zion with its canyon river and spring waterfalls splashing down the great red cliffs. I could live at Zion and never feel sad about leaving civilization behind.

"What if we take vans?" Dr. Stillwell asked me across his office desk February 11, 2011. "How far is it to the Grand Canyon and back?"

I sat in my chair by his door, journal in my lap. "That's a lot

---

1   See "What Are Bryozoans, Anyway?" in the Appendix.

of time for people to be together in a van," I grimaced. "If we stop in the Ozarks like you wanted, it's more than 2300 miles one-way. That's four days out and three days back at least. Unless you want to drive 20 hours a day."

We were traveling out west in May. We were going if I had to drum up a teleporter to flash us there. Dr. Stillwell was working it out with the dean, and between the department and some funds of his own, he figured we students would only have to pay $500 out-of-pocket, which was ridiculously cheap.

Since bryozoans were on hold, Dr. Stillwell had me work on the class manual for the Grand Canyon trip. This satisfied that whole "Ooh ooh! What cool geology does Bryce Canyon have!" side of me, the Dr. Livingstone get-up-at-2am-and-go-exploring side of me. It also meant that I was sitting inside the door of Dr. Stillwell's office every Friday afternoon. Dr. Stillwell had a lot on his plate, but on Fridays he cracked open a few minutes of time to sit with me and plan a magnificent trip.

On February 11th, I sat in my chair as Dr. Stillwell counted out loud. "One… two…three days at the Grand Canyon, two days in Zion and two days in Bryce."

"And you wanted to go over Escalante," I reminded him as Flash stepped through the door.

"Dr. Gurden, what if we make the trip 14 days long?" Dr. Stillwell asked.

Flash stared at Dr. Stillwell with less than great enthusiasm. "What happened to nine days?"

"We're thinking about driving instead of flying," I said.

Flash shook his head. "My wife is pregnant. When I suggested we'd be gone 10 days, she freaked out. Ten days is exponentially more than nine days."

"So two weeks is out of the question," I nodded at Dr. Stillwell. It was plain ridiculous to drive.

Flash had become Dr. Stillwell's partner-in-crime on geology trips. He'd trekked with us up Seneca Rocks the previous fall, and he'd been crimped into the Grand Canyon trip as well. Not kicking and screaming, though, because those were fun times.

Still. "A week on the road won't be fun," I urged Dr. Stillwell to reconsider. "It'll be okay if we fly. I saw plane tickets for $300. That will leave $200 per person for minivans and food and gas." We'd all be sleeping on the ground in tents, not staying in hotels.

Dr. Stillwell frowned at me with less than great enthusiasm. He kept trying to avoid putting 10 people on an airplane.

He sighed. "We'll see how many people come up with the funds to go."

The next Friday was an important day, a day that demonstrated just how far I'd journeyed in my relationships with these men. That particular Friday, February 18th, Dr. Zenith blocked my way when I arrived at Dr. Stillwell's office, and he closed the door in my face.

That was the sort of thing Dr. Zenith would do. He would close the door in my face.

I did not expect, "Excuse me Amy Joy," from him, so I smiled to myself and wandered into Dr. Gurden's office to wait. Flash and I were buddies. Flash never treated me rudely.

"We need to figure out how to stop them," the young biophysicist said as I sat down.

"Stop who?" I asked.

"The stink bugs," Flash grimaced at his computer. "The first documented stink bug was seen in Allentown, Pennsylvania in 2001, and after just 10 years they're a full-blown plague." He looked at me. "Did you know they can produce three generations each year?[2] And they hibernate, so they can live three or four years!"

Halyomorpha halys, also known as the brown marmorated stink bug, had become a gross problem in 2010 and 2011. Gross as in "large" and gross as in, "Ewww." Native to Southeast Asia, they were an invasive species in Pennsylvania and surrounding states, and they chewed up orchard fruit and other agricultural crops with few native predators ready to take them on. They'd hibernated in my cabin's loft and I'd been shaking them out of my sweaters - and it wasn't even spring yet.

---

2 According to the EPA, they produce one or two generations per year, but up to five in warm climates.

"I was at the 7-11 on Highway 45 getting gas last year," Flash continued, engrossed in his animosity. "I looked up, and the hood over the gas pump was just crawling. Just crawling! It was… it was…" he searched for the right word. "It was unbelievable. They destroyed my garden last year, laying their eggs under the leaves. They carpeted the side of my house. I had to go out with a Shop Vac to vacuum them up, and it didn't even make a dent!"

I watched him scroll down his computer screen.

"So, are you planning… to do something?"

"I'm going to experiment on them and dissect their eyes and remove the pigment and find out how they see. I want to design a way to lure them into traps."

Wow. No mercy! Dr. Gurden had declared war on the stink bugs. He could do it, too. He was a bright and inventive biophysicist. We sat together, brainstorming on possible counter measures to take against the invaders. We could confuse them with blinking lights. We could use pheromones to draw them into stink bug hotels from which they'd never escape!

After awhile, we heard the door shut next-door. "Dr. Zenith is walking down the hall," Flash nodded at me.

I hopped up to make another try at Dr. Stillwell's office, and Dr. Zenith blocked me as I tried to enter the hall. "Why was my name called?"

"Oh," I shrugged. "Dr. Gurden was just narrating the goings on in the hall. Dr. Stillwell asked me to come by at 3:00."

"What that man needs are some cookies!" Dr. Zenith declared. "He really wants some cookies. My wife's got cupcakes for me. I'm outta here." With that, the astrophysicist marched down to his office.

I thanked Flash and walked to Dr. Stillwell's doorway.

"Come in! Come in!" Dr. Stillwell urged me. I grabbed a handful of chewy candy from the candy pumpkin on his bookshelf and settled into my seat. There I saw the furrowed eyebrows and dark demeanor that told me Dr. Stillwell's meeting with Dr. Zenith had covered some less-than-pleasant subjects.

"Are you having a bad day?" I greeted the geology professor with compassion.

"Oh," he said. "There are just people who promise things they can't deliver, and now I have to be the bad guy and inform everybody that I cannot give what was promised." He wanted to go home and turn on some hockey. "Watching people beat each other down on the ice sounds very satisfying right now."

I didn't know how to help. I offered, "Can I be your enforcer?"

"If you had your PhD, I'd set you on it. But you don't have your PhD *yet*." He said, "yet" with encouragement, full of faith I'd get there.

He handed me a geology book on national parks. We were planning trips! The troubles of the day would soon be dusted away behind visions of red cliffs and pink stone spires. "The human race is worth saving," I reminded Dr. Stillwell as I opened the book. We were not stink bugs requiring eradication, no matter how he felt at this moment.

"Maybe one or two," he grimaced.

"Well, if only one or two are worth saving," I shrugged, "then none of us are."

"Maybe one or two," he insisted.

I chewed on candy and turned book pages. After a few minutes, we heard a strange buzzing in the hallway. The geologist and I looked up from our work as Flash appeared outside the open door, driving a sizeable remote-controlled airplane. Sunlight shone into the hallway from large windows, and Flash needed to take his big white plane outside for a test flight. The white drone taxied down the tiled floor, about five feet from wingtip to wingtip, ready for espionage.

"Is this your newest weapon in the war on the stink bugs?"

Flash didn't answer me. He'd encountered a problem with the remote control, and the drone had stopped. Flash pressed full-forward on the levers, but the plane wasn't going anywhere. Puzzled, he bent over and fiddled with it.

Dr. Stillwell leaned on his desk to watch, readying himself to offer a suggestion. Flash crouched and pushed a button.

BAM! The plane smashed full force into the wall.

Flash's eyes went wide.

Dr. Stillwell exploded with laugher. "Hahahhahahaha!!" His delight echoed all through the hall, a welcome release from the tensions that had hounded his day.

Dr. Zenith shot from his office and marched our way. "That was AWESOME!" He immediately whipped out his phone and took pictures of Flash and the wreck.

Dr. Stillwell kept laughing loudly. Flash bent to collect the broken plane parts scattered on the floor while Dr. Zenith danced around, full of witty Dr. Zenith commentary.

And I realized something wonderful.

I realized that a steep mountain pass had been conquered. Flash had welcomed my chatting with him about stink bugs. Dr. Stillwell had invited me into his office despite his aggravation and desire to watch hockey violence. Dr. Zenith didn't even question my presence in Dr. Stillwell's doorway. Nobody treated me like an annoying little sister or told me to go away. I had become a member of the group, part of the tapestry. I belonged.

I kept thinking about the passage God had given me in November. Through Isaiah 30:20, He'd told me that my teachers would not be placed in a corner anymore, that I'd see my teachers. I'd read that verse in November, and there it was, living itself out in front of me. I loved these guys. I didn't want to lose them.

And then there was Isaiah 30:21:

Your ears shall hear a word behind you, saying, "This is the way, walk in it," Whenever you turn to the right hand Or whenever you turn to the left.

The whole passage had been promised to me - the bread of adversity and the water of affliction, seeing my teachers, and careful direction when I needed it. Sure enough, along with those friendships, the Lord's careful direction came to life that semester more than any other time in my history.

CHAPTER 7

STINKY BUGS

"If I died and went to the bad place, there would *still* be cockroaches."

-Joe David

Dr. Stillwell treated me kindly. My atheist professor forgave me for being unorthodox, and I appreciated that. I forgave him for being hostile to religion. I figured his beliefs weren't between him and me, and God was big enough to handle Dr. Paul Stillwell all by Himself. He didn't need my help.

Still, it didn't take long for me to infuriate Dr. S. that Spring 2011 semester.

I didn't do it on purpose. I read a lot, and every once in awhile I came across articles I thought might interest certain professors. I didn't plague their in-boxes, but Dr. Bob might be interested in an article about nanotech treatment of cancer, and Dr. Stillwell might like to hear that feathered dinosaurs were found in China. One day that spring, I shot Dr. S. an email containing an article that declared, "13% of H.S. Biology Teachers Advocate Creationism in Class."[1] He despised creationists with a palpable passion, but I wanted to clarify the "13%" statement, which I thought was misleading.

I don't know if my mistake was to send the article in the first place or if it was the short note that I wrote with it. The article treated

---

1 Welsh, J. (2011, January 27). 13% of H.S. Biology Teachers Advocate Creationism in Class. *LiveScience*, retrieved April 18, 2015, from http://www.livescience.com/11656-13-biology-teachers-advocate-creationism-class.html.

Intelligent Design Theory and Creationism as the same thing, and I saw a distinction. Creationists tend to start with religious texts and move from there to the physical world, while I.D. theorists credit a Designer with the engineering they see in nature, moving from the physical world to belief. Creationists tend to have religious motivations, while those who promote Intelligent Design can include agnostics and even atheists who think Earth was seeded by aliens. I commented on this distinction in my email and clicked the "send" button, innocently, not suspecting the wrath I'd just planted in the heart of my favorite geology professor. I went about my day, which turned out unpleasant on many fronts.

One thing after another banged my emotional shins that afternoon. I'd been dealing with horrible loneliness all weekend. I still suffered from grief 18 months after Randy's death, and that weekend had proved particularly painful in a way that leaked into Monday. That afternoon we watched another physics teacher candidate session, and the guy wasn't even very good, but I still managed to say something stupid that I wanted to take back as soon as it left my mouth. It was just a lousy day.

Basically, things were already rotten when I sent that little note, and because I sent it, Dr. Stillwell's blood pressure spiked. On the outside, the professor remained friendly, but … tense. He didn't hide his feelings well, and it wasn't long before his aggravations burst through their half-closed door.

In a moment of levity, I joked with Flash and Matthew in the hallway. We let the afternoon sun shine on us from the tall windows and talked about stink bugs as Dr. Stillwell walked up and joined us.

"The stink bugs have no natural predators here," Flash said. "That's one of the biggest problems."

"What are their predators in Japan?" I asked.

"Wasps, I think," Flash said.

"We're their predators," Matthew said. "We can suck them up with a wet vac."

"Yeah," I said, "but when you vacuum up stinkbugs, it fills the room with their odor. I think they smell green."

Dr. Stillwell frowned. "I hate to think what kind of salads you eat, if you think stink bugs smell 'green.'"

"What do you think is the stink bug's purpose?" Matthew asked. "What do they do?"

"I don't know," I said. "They must have some value in their native ecosystem."

"Well, after all, they were intelligently designed," Dr. Stillwell said in jest: an acidic, bitter, emotional form of jest. "Some intelligent designer *designed* them."

That's how he started, and he let himself vent for half a minute. He went off, hammering the term "intelligent *design*" over and over. I stopped hearing his words, because I was focused on the hardness of his eyes and the irritation in his voice.

The rest of us tried to joke a bit about the stink bug invaders. "Well, they were here before we were." It didn't do much good. Dr. Stillwell's tension filled the hallway, and soon Matthew and I retreated, leaving the geology professor to bubble in his own cauldron.

We walked to the elevator and let the door close on us.

"Will you pray with me?" I slid to the elevator floor. "It's been an awful day, and there's such a heaviness everywhere."

"Yeah," Matthew agreed. It was affecting everybody.

Chia Pet math genius Matthew and I had become prayer partners during the previous fall. He fasted on Thursdays, so I said I'd join him. Every Thursday we got together and prayed for each other and our friends and professors at school. We prayed for our families and anything that came to mind. Then, we ate a meal Thursday evening.

That day, we retreated to the safety and quiet of the elevator and prayed together.

I felt a lot better after that, but I still argued with Dr. Stillwell in the mirror that evening.

Dr. Stillwell's outburst wasn't reasonable. His anger was petulant and harsh, and it had nothing to do with me. He took it out on me, and I didn't appreciate it. I mean, he had a right to be skeptical about the value of the brown marmorated stink bug. I couldn't see anything good about the wretched insect either – not for us humans, anyway. But that wasn't the issue. There was nothing balanced about his attitude.

Didn't I know that the concept of Intelligent Design was frowned upon in the halls of academia? Of course I did. That wasn't even the point. The point was that Dr. Stillwell didn't bother to talk to me about it. He just got all emotional and bad tempered, going to the worst place in his mind without even asking me questions. He constantly did that!

It was obvious the next day that Dr. S. felt a little sorry and wanted to make up with me when he wandered into Dr. Gurden's physics class. He saw me across the room and made a frowny face to mimic my own serious face, then he paused by my seat to check the physics equations on my paper.

After class, I stopped into his office to get a weather forecast, since he was my nearest meteorologist. "Why are you being so quiet, Amy Joy?" he asked me gently from his desk. "You're being awfully quiet for you."

I didn't say it out loud, but I thought, "You purposely bashed me about the head and shoulders with your Intelligent Design nastiness, and now you're trying to make nice, and I'm not over it."

I was willing to forgive him, but I had to get some things straight with him first, and I didn't have time at that moment. I said, "Let me talk to you about it later. What's the weather going to be like tonight?" That's what I needed to know. I had to drive an hour to school on windy roads in the February cold.

He said, "Well, it's probably going to rain tonight, but it might freeze. You drive carefully on the ice, okay?"

"Okay."

Dr. Stillwell had gotten all miffed in the hallway on Monday. On Tuesday, he'd tried to be nice. On Wednesday, he and I had a morning meeting to plot our Grand Canyon stuff. I entered his office at 10:00 a.m. and settled into my seat, but he didn't even bother to hand me the geology book. We needed to get some things worked out.

"Listen," I got to the point. "You've been driving me crazy ever since I answered the Noah's Flood question last year. You're always suspecting me of trying to prove religious things, and I can't ask questions about catastrophes without your eyeing me with suspicion. I'm not trying to prove anything. I'm just asking you a question."

"I have?" he asked.

His question wasn't really a question. His "I have?" was a deflection. It was a tool he used to avoid having to say anything.

"Yes!" I glared at him, irritated with the false innocence. "You were uncomfortable around me for weeks until I told you I'd made casts of the *Archaeopteryx*."

"I remember that," he said. "That was neat."

"And when I told you that, this huge weight lifted off you."

He deflected again with, "Did it?"

"Yes. And when I sent you that email earlier this week, I wasn't sending the article to pick a fight. I just sent it because you like to fuss about our education system. So, then I had to endure your getting all tense and being nasty."

"Was I nasty?"

"You were irritated and tense and attacking Intelligent Design at me."

"Was I?"

"Don't pretend you don't know what I'm talking about!" I looked him in the eyes earnestly. "I need you to trust me," I said. Please. Please, Dr. Stillwell. "I need you to trust that I'm doing my best to take the data and base my conclusions on all of it. I am not using it to try to prove what I already believe. Okay?"

The false innocence left his face, and he nodded in honest agreement. "Okay."

I settled back in my chair, giving him the floor.

He did speak his mind then, which is what I've always wanted from him. I just wanted him to say what he was thinking without going into attack mode.

"The problem I have with the Intelligent Design people is that they have an agenda," he said. "They tend to be the village idiots who ignore facts that don't suit them. They have the things they want to promote, and they already have their minds set. They found that Creationism didn't go over well in the courts, so now they've tried to dress up Creationism in scientific terms and call it 'Intelligent Design,' but it's all the same people."

I liked talking to Dr. Stillwell about these things, because he gave me a view on what many university academics honestly believed to be true. His ideas were contrary to my own experiences, so I forgot that many people thought like he did. It was important to be aware of his perspective. Still, he was not judging the situation with enough information, because the creationists and the Intelligent Design folks are not a homogenous group of people.

We could just as easily accuse every evolutionist of wanting to destroy the faith of millions of children. Certainly some evolutionists like Richard Dawkins want to do just that. Long ago, Dawkins moved past pure science into the realm of philosophy, and he's gone on a crusade to crush religious belief. Still, it's not right to assume that every evolutionist professor wants to drag students into Hell. It's just not true. Some people are evolutionists because they (crazy, I know) think the evidence points to the grand theory of microbes-to-man evolution.

Creationists are often regarded as religious fanatics who pick and choose the data they like, the data that fits their view that God made the earth and everything in it. Perhaps some have moved into the Intelligent Design category in an effort to escape the stigma associated with "creationist." Perhaps some are village idiots. However, many are *not*. In my experience, I've found a wide spectrum of intelligence and education among those who question the "Evolution did it all" explanation of our origins.

I once met former NASA engineer William Overn at a gathering in a friend's living room in Albert Lea, Minnesota. It wasn't until later that I learned that Overn had been involved in developing the computer memory systems for the Mariner IV probe, which took the first pictures of Mars from space back in the 1960s. I didn't know he'd worked for NASA! As far as I was concerned, he was some affable old guy involved in computer programming, willing to chat with me while I munched on snacks. He told me that early in his career, the computers kept crashing, and the programmers had to repeatedly rewrite the code to get the computers up and going again. Finally, the project manager came in, nearly pulling his hair out, and he said, "We need to make these computers pick themselves up by their own bootstraps!" According to Bill Overn, that's where the term "to boot up" a computer came from.

It turned out that Overn was one of the founders of the Twin Cities Creation Science Association, a completely and utterly young-earth creationist organization. He didn't talk to me about a young earth. He talked to me about the history of computer science, which I thought was fun. Nothing about him said "village idiot."

I've met a wide variety of creationists over the years here and there, including several professional geologists. I once even met old Henry Morris and John Whitcomb, creationist authors of *The Genesis Flood* (1961).

Fun story about John Whitcomb. Dr. Whitcomb asked me if I could whistle through my hands, and I said I could. Back in the eighth grade I'd spent three days struggling to learn this odd skill, so I made a box with my hands and whistled for him, and he made a box and whistled together with me. There's a video somewhere of John Whitcomb and me playing "Jesus Loves Me" on our hands. He also talked quite a bit about entropy and the Second Law of Thermodynamics, which I found interesting.

Creationists take the words in Genesis 1 as a literal description of Creation Week, and they see evidence in the world to support that belief. They don't hide behind Intelligent Design; the Bible as God's Word comes first.

The Intelligent Design folks are a different group than the creationists. I haven't met those in the Intelligent Design community, but I've been reading their books, and nothing about their writings seems bent on proving the first chapter of Genesis as history.

Whether people agree with the Intelligent Design folks or not, it's foolish to dismiss them as creationists in disguise. Most seem content with the Big Bang explanation for the origin of the universe and a 4.6 billion-year age for Earth - both of which are rejected by young-earth creationists. I've found the I.D. proponents to be good writers who offer careful, thorough analyses of the troubles involved in evolving biological systems from scratch.

Dr. Stillwell's opinions didn't match my observations.

There is a long list of professional scientists who think evolution is insufficient to explain life as we know it.[2] It was arrogant and unjust for Dr. Stillwell to put everybody in the same box. He could not possibly know the motivations of every creationist or every Intelligent Design proponent. I trust there are plenty of I.D. theorists who take up their position because they (crazy, I know) think the evidence points to a Designer.

The general scientific community and the media have a shamefully bad habit of accusing one group or another of being "anti-science." That's one of my pet peeves. It's troll talk. It's cheap and weak. If a position is strong, then its proponents should simply lay out the facts and leave off the emotional ad-hominem attacks. None of us are in a position to accuse another group of being "anti-science" simply because we disagree with their position.

"You're a big stupid head!"

Honest to goodness, everybody stop with the name-calling. Seriously! We all know it's fun and emotionally gratifying, but we're big people. We can tie our own shoes and everything.

I sat across from my dear geology professor and thought about what he'd said.

"I took biology at a Jesuit high school," I told him. "My biology teacher faithfully tried to teach us evidences for evolution. 'Ontogeny recapitulates phylogeny' is permanently burned into my mind," I smiled.

---

2   A list of professional scientists have signed the statement, "We are skeptical of claims for the ability of random mutation and natural selection to account for the complexity of life. Careful examination of the evidence for Darwinian theory should be encouraged." As of this writing, the list includes 22 pages filled with the names of degreed scientists, more than 800 names. See A Scientific Dissent From Darwinism, (2015, February 1), Retrieved April 30, 2015, from http://www.dissentfromdarwin.org/

Dr. Stillwell offered a smile in return.

I continued, "But, the more I studied, the more I thought, 'Wow. This is amazing. This happened by unguided natural causes? That's... that's ridiculous.'"

I truly believe that if I had been raised in an atheist home, the study of biology might have led me to believe in God. It amazed me that anybody could conclude for a second that life engineered itself. Nothing engineers itself ... unless it's already been pre-engineered to do so.

Dr. Stillwell is pretty smart. I'm pretty smart too, and I think we both care about what is actually true. That gives us good potential as a team, because we can pull out both honest sides of any argument.

"I put myself in a vulnerable position with you," I reminded Dr. Stillwell. "I open up to you, and I say things to you that should get me into trouble. Like - I tell you I think that God raised me. One day I'm going to need you to write letters of recommendation for me, but then I tell you something like *that-*"

I rolled my eyes at myself. I was asking for it.

"No, but I think you're right," he said calmly. "You and I both have a love of nature. And retreating into the outdoors is akin to my deity, if I have one at all. Your belief in God offered you something you needed. I'd go out under the cedars there on the Olympic Peninsula when I was a child, or I'd stare up at the redwoods, and I think that's where the idea of churches came from."

I got that. Dr. Stillwell had something there about the architecture of church buildings. It's like heaven for me out in the woods under the trees.

"I do love nature," I told him. "But my God is more than nature itself. He's the wonderfully clever God who created it all. He's greater than the world He made. It's so fun to be able to say, 'Wow, God. You're so neat.'"

Dr. Stillwell smiled at that. "Way to go. Thumbs up."

I knew he was mostly trying to make peace with me, but I gladly accepted it. During the next half hour, I told my atheist professor about some of the miracles I'd experienced in my life. I told him about the vision Randy had of the brutal horse accident I'd suffered when I was 15 and explained that Randy had been able to accurately describe an accident he hadn't seen physically.

Dr. Stillwell didn't offer any skepticism. He gently listened until I'd finished. Then he gazed at me steadily and quoted *Hamlet*.

"There are more things in Heaven and Earth, Horatio, than are dreamt of in your philosophy."

It's true, Horatio. There are many more things.

Dr. Stillwell and I talked. It was a good, calm, cordial conversation, and I was grateful for it. It was okay for us to disagree about things. That was okay. I just wanted to do it as friends who respected one another.

CHAPTER 8

DESKS IN THE BATHROOM

AJ: They say, 'No. Don't you DARE give Amy Joy coffee!

Dr. S: We don't let her have high fructose corn syrup, let alone anything with caffeine.

My teachers have always been important people to me. When I was a little kid, it was the teachers who got my jokes. I'd say something funny and wait for the other kids to titter, and none of them did. Dead silence. Total rejection. I didn't find out until much later that the teachers had been stifling their own laughter.

I've always been something of a nerd. It's hard to make friends when you're a nerd, because nobody sees who you are. They don't get your jokes, and they don't understand how you think, and you're not cool. They're all absorbed in their own insecurities, and they don't think you're the best person to make them look good.

Being a nerd is lonely, unless you can find other nerds to share your time. And even then, I was less cool than the other nerds because I was both a nerd and *poor*... and maybe I smelled bad. At one point in high school, I had to walk three miles just to catch the bus to school because I couldn't depend on my mom's van to start in the morning. My schoolmates' parents were bankers and lawyers and surgeons. In a prep school full of rich kids, I felt like an unwanted. A leftover.

As a child, I learned that the people who appreciated me were the teachers. The teachers didn't care that my clothes came from a thrift store. They listened to my stories. They thought I was cool. One of my favorite teachers of all time was English guru Mr. Mike Carroll, and we're friends to this day.

Mr. Carroll started an Odyssey of the Mind team my sophomore year of high school, and that's where I found my niche. Odyssey of the Mind fit me perfectly, me and the other creative nerds in my world, nerds much richer and cooler and better smelling than I was. Tiffany M. and I sat on the cold tiles of the school hall late one night, attaching a balloon diaphragm to a PVC-pipe pterodactyl along with two clothes pins. Enya's "Orinoco Flow" echoed soothingly through the hallway. I blew into the balloon to fill it, then captured its latex mouth tightly between the two clothes pins. I released the lips of the latex, and our pterodactyl squealed gloriously.

Odyssey of the Mind promotes creative problem solving. It's about taking 10 minutes to build a device that can shoot a ping pong ball through a hoop across the room using only toilet paper rolls, paper clips and troll dolls. It's about developing a 10-minute play that presents the problems each team must solve for the year – the team's competitive answer to the year's creative challenge.

My senior year of high school, our Odyssey of the Mind team built five robot dinosaurs that moved in a variety of mechanical and electrical and pneumatic ways. They were awesome. Our team produced an ankylosaurus that rolled across the floor and a theropod that used bicycle gears and chains to move its theropod feet and its theropod arms. Our squealing pterodactyl swooped and our giant backdrop loomed overhead with massive conifers made from carpet padding that I'd found in an industrial-sized dumpster off Trent Avenue. They were the best dinosaurs at the regional competition, and we'd have won except that somebody (I'll take responsibility) forgot to plug in the power strip before our skit started.

Still, I won The Far Side award that year as MVP of our Odyssey of the Mind club. Mr. Carroll presented me with a chimera of a trophy sporting a tennis player and a bowler and a boomerang, all sticking out in odd directions at different heights. Best of all, the trophy was topped with the back end of a horse, and we all know

Figure 3: The top of The Far Side Award that Mr. Carroll gave me. It was the very first Far Side Award ever presented by Mr. Carroll, and obviously a precious item. I keep it on my piano. (The upside-down OM is symbolic of creativity and intends no disrespect.)

what that means. Apparently, trophy shops can carry horse rumps as a topping option. I am supremely proud of that thing, and you can't have it.

Finally, there came the day we had to clean up our Odyssey of the Mind work room. We had littered our small dungeon in the school basement with junk and wood and metal and tools and supplies we'd used in making the play sets for our O.M. competitions. I was graduating. It would be my last hoorah with my team.

"All right," Mr. Carroll told us at our last meeting for the season. "We have a job on Friday afternoon. We have to clean up the basement, and everybody had better show up."

You know Mr. Carroll. His blonde hair is a bit wild, and he looks like Gene Wilder's version of Willy Wonka without the purple hat. He orders chimera trophies topped with horse rumps. Insolent students in his classes can expect to fill buckets one cautiously balanced spoonful at a time from the water fountain at the end of the hall. Yeah, you know him.

I interjected, "Um. I have a babysitting job after school. Would it be okay for me to bring the kids to the cleanup party?"

Mr. Carroll frowned at me. "You have to come. You don't get to skip out while the rest of the team cleans."

"No," I clarified, "I don't want to skip out. I want to help clean the basement, but I have this job. So, could I bring the kids or maybe help clean at another time?"

"You can't get out of it," Mr. Carroll said - deadpan - having himself a bit of fun.

"I don't want to get out of it," I said. "I just want to know-"

"You have to help," he interrupted. "What's wrong with you, trying to get out of work when everybody on the team has to help?" There was no mouth smile, but his eyes glinted and danced in self-entertainment. Amy Joy is the joke of the day. Ha ha ha ha ha!

The other team members gleefully joined in. "Yeah! You can't get out of it. You have to help clean too!" They kept teasing me, hollering behind me as I grabbed up my backpack and left the room.

I closed the heavy wooden door and marched down the hall tiles, revenge bubbling all sulfur hot springs in my chest. They were only teasing me because I *wasn't* lazy, but they were amusing themselves at my expense. I would get back at Mr. Carroll, that poo head. I had to think about how to handle it.

I could refuse to help at all!

No. No, I couldn't do that.

I could just show up with little Eric and Anna, my after-school wards, and clean with everybody else.

No! I would surely have my revenge!

I could come in and clean early…

What to do. What to…

I stopped. Oh. Ohhhh…My eyes widened.

"I'll just clean the whole thing myself!"

Bwahahaha! What a delightfully rotten thing to do!

I sneaked into the basement the Thursday night before the big cleanup day and overhauled that whole work area. I organized all the boards and carpet remnants and glue and tape and chicken wire and paint and plaster and metal parts and tools. I picked up all the screws and nails and bolts and nuts and washers off the floor. I put things away if I knew where they went, and I stacked things neatly when I didn't. Then I swept the whole thing. My dear friend Ed Goodman showed up that evening with a Subway sandwich for me, offering me nutrition in the middle of my labor. God bless Ed.

The unsuspecting Mr. Carroll had no mercy the next day. He continued to hammer me at the mid-morning break, warning me, "You have to come. You can't be lazy and leave everybody else to do it all!"

I grumbled and shrugged him off. Heh heh heh.

Then, I forgot about it. I trotted off to my classes and got involved in calculus and physics and French. I walked into Mr. Carroll's room after school and plopped my books on his window sill like always. I turned to my wild-haired English teacher, and the man's whole attitude had changed. His shoulders slumped like a Tigger who'd lost his bounce. He gazed from under his blonde eyebrows. "Did you… did you clean the basement?" he asked weakly.

Ha ha! "Yes, I did!"

His contorted body reflected the internal agony I'd caused him. Shame racked him in a very pained, self-flagellating way. I could picture his quick-stepping down the stairs into the basement to plan his attack on the mess below only to discover the clean-swept room where a wreck should be. I felt the little stool getting knocked out from under his feet. "OhhhhHHH!" he groaned. "Talk about killing somebody with kindness!"

I glowed to the top of my vicious and cruel head. Sweet sweet revenge.

This all ties back to the spring 2011 semester and Dr. Zenith, by the way. But, there's a second part to the story that's important.

Two years earlier, Mr. Carroll and 11th grader Sean M. stood out by Mr. Cossette's little green car in the teachers' parking lot, contemplating various plots of evil against it. Mr. Cossette had recently joined the faculty as the new debate teacher, and few guards were lined up to protect his poor green car.

Earlier that year, Mr. Carroll and student assistants had removed Mr. Cossette's desk from his classroom and carried it away up the stairs, where they shoved it into the girl's bathroom. They placed all the items that once sat on top of Mr. Cossette's desk onto the floor of his classroom in exactly the same locations those items had occupied on his desk top. The desk remained hidden in the girl's bathroom for nearly a week. Nobody told Mr. Cossette where it was. Nobody.

I anticipated great mischief toward Mr. Cossette's car.

I approached Mr. Carroll and Sean on my way to the student parking lot, both curious and trusted. Mr. Carroll stood, arms folded, puzzling with Sean over possible deviltries.

"We could set it up on cinder blocks, a quarter-inch off the ground," Mr. Carroll told Sean. "When he gets in, the wheels will spin, but he won't go anywhere."

"That's great. And there's a kind of spray paint we can get," Sean suggested. "It wipes off, but he won't know that."

Just then, Mr. Cossette himself exited the building and spied Thing 1 and Thing 2 standing by his car.

The debate teacher asked with justified suspicion, "What…are you doing?"

"Oh nothing. Nothing at all. We weren't planning anything." Mr. Carroll and Sean sauntered off, freely radiating their guilt.

By, the way. Sean was my foster brother. Sean's younger sister Megan was my best friend in high school, and I lived with Sean and Megan that school year while my mother got settled in northern Idaho. Their Irish Catholic father served as the city's chief of police, and he also served legally as my foster father. Little details.

Not many days later, I had to remain after school for some evening event, so I huddled in Mr. Carroll's classroom where other students played chess or did homework. While our faithful Odyssey of the Mind coach enjoyed a dice game with students on the other side of the room, I labored at his desk, aggravated by cabin fever.

I sat in Mr. Carroll's own chair, writing de-stressing poetry about exploding cows.

I finally decided to go for a walk. I'd breathe fresh air and free bulging frustrations from the confines of my soul. The fresh spring air. The sun. These were good things. I headed out the heavy wooden door and down the long hallway past my locker, out the same door that Mr. Cossette had exited a few days prior. The teachers' parking lot ran the full length of the school in those days. As I pushed out the door, I noticed that Mr. Cossette's parking space sat open and empty, but Mr. Carroll's Toyota Celica sat right there, exposed to all the world. Mr. Carroll's car! My mind shot to the spot where I'd left my exploding cow poetry. His keys. He'd left his keys right on the desk.

Ohmygosh.

This particular week I had a bit of a grudge against my O.M. coach. On the day that he and Sean had plotted evil against Mr. Cossette's car, Mr. Carroll had snubbed my Odyssey of the Mind team at the school award ceremony. We'd won at regionals, and he'd ignored us while celebrating the team that took 2nd place at state. It was only a small grudge, but combined with Mr. Carroll's own practical joke plans and my vision of his keys sitting openly on his desk… I had to do something. Bwahahah!

I wanted to be able to move the car, but I didn't want to do anything really awful to it. Actually, no. I did want to. I wanted to put that Toyota Celica on the school roof. That's what I wanted to do. But, since I couldn't do that, I tried to think of something relatively obnoxious and simple.

We could drive his silver car into the tennis courts (Mr. Carroll also coached tennis) and fill it with tennis balls.

We could tie a big bow on it and paint big woman's lips on the hood and long-lashed eyes on the windshield.

Or heave it up and rest it on cinder blocks one-quarter inch off the ground.

My biggest problem was that Mr. Carroll's car had a clutch, and at 15-years-old I still had little practice with manual transmissions.

Nothing actually possible came to me, not when I couldn't drive the thing, and I only had a few minutes to think.

I walked calmly back to the classroom, dropped into Mr. Carroll's chair at his desk, grasped the keys so they didn't jingle, and slid them gently across the wood and into my shirt. I walked calmly back into the hall and there began the difficult task of finding somebody who could drive a stick shift.

Which was nobody. Because anybody stuck at school obviously had no wheels. We were all underclassmen.

I still wanted to put the stupid car on the roof. If we were seniors with skills and muscle and time to plan, we could have done it. But, we were skinny kids with ten minutes to figure out how to drive a clutch.

Erik Esvelt handled it in the end. A freshman named Deanna and I were able to get Mr. Carroll's car to coast down to the curb. Erik managed to back the car onto the lawn under the tree on that far side of the building. This had the simple effect of hiding Mr. Carroll's car from view when he walked out to go home. It was the best we knew how to do on short notice.

I wrote an unsigned note that verbally laughed at the English teacher and dropped it onto the driver's seat. I figured he'd know my writing. I strolled back into the classroom and smoothly returned the keys to their spot on the desktop - just about the time everybody was getting up to leave.

I said goodbye to my retreating peers and waited for Mr. Carroll to walk out. As soon as he started down the hall, I slipped out the other school door and peeked around the bushes, waiting for him to emerge at the end of the parking lot.

After a minute, the door opened. The victim appeared and strolled down the sidewalk.

He reached the space where his car should be, and I expected a little excitement. I expected a bit of jumping up and down and yelling. Nothing. The teacher looked this way and that, then he wandered a bit farther and spied his car under the tree. He got into it and drove away.

That was it. Major anticlimax! Such a disappointment.

The next morning, I dashed into Mr. Carroll's room to grab a book, but he made no comment about his car. I figured he'd say something about my trespass. Nope.

He didn't seem rumpled about it at lunchtime either.

Darn. We should have foamed up his hood with a big whipped cream mohawk and parked it on the principal's lawn. We should have stuck a statue of St. Ignatius of Loyola behind the wheel, complete with sunglasses.

After school, I wandered into the English teacher's room to meet Sean and Megan for my ride home, and Mr. Carroll stopped me at the door. He frowned, puzzled. "Did you… did you move my car yesterday?"

What? "Of course I moved your car! I thought you knew I moved your car!"

Willy Wonka's blue eyes widened. He ordered everybody else to leave his room. "All right!" he raised his voice. "I need everybody out. Exit the premises. Let's go!"

Oh wow. This was a reaction!

I'd *wanted* Mr. Carroll to jump up and down and make noise. He didn't bellow, though. As soon as the other students left and we had the classroom to ourselves, he did something much scarier. He got super quiet. Super, rage-hiding controlled and soft spoken.

Perhaps I'd made an error.

"Who helped you move the car?" Mr. Carroll asked in that soft, dangerous voice.

"I'm not going to tell you that." How could he ask such a thing? There were codes of honor involved here.

"Who helped you move my car?" he insisted.

"I'm not telling you."

Mr. Cossette happened to enter Mr. Carroll's doorway at that point.

"Hello, Mr. Cossette!" I cheerily demonstrated my lack of interest in Mr. Carroll's dangerously quiet disposition. "How's your day going!"

Mr. Cossette's own disposition glowed with amusement.

When he left, Mr. Carroll got mean. "If you don't tell me who helped you, I'm going to talk to Mr. Tchida."

Ohhh. That was low, involving the vice-principal in all of this. Mr. Carroll could dole out J.U.G. (Justice Under God) himself. He could make me clean his floor with a one-inch-square piece of felt. Or repeatedly dig and refill a hole in the courtyard. Or fill a bucket one spoonful at a time from the water fountain at the end of the hall. This was between him and me. It was low of him to threaten me with the authorities.

"Why should I have needed help!"

"Because you couldn't have picked up my car and moved it on your own. Who helped you?"

"I didn't pick up your car," I corrected the naive man. "I stole your keys."

"You took my keys!" I thought his eyeballs would pop from his head and hit me in the face. I was charmed and alarmed at the same time.

"Don't you ever ever ever do that again! Never again. Promise me."

"Okay."

He ordered me to leave while he considered the proper retribution for my crime.

I grabbed my backpack and slumped across the hall to an empty French classroom. There I sat and considered the error of my ways.

Mr. Carroll's anger didn't bother me at all. He deserved it, wicked man. However, I considered the possibility that I had damaged my friendship with him, and that concerned me deeply. I depended on his friendship, and this whole silly thing had bothered him in a different way than I'd thought it would. (I joked about moving his car years later, and he instantly tensed up. "Don't you dare.")

I felt bad. I worried that I'd messed up things with him forever.

Sean found me grieving in the French classroom five minutes later. He offered me a high five.

"That was awesome!" Sean said. "When I went into his room

after school today, he was all, 'Hey Sean! How'd you get my keys Sean? How'd you get my keys?' Then Megan went in, and he got all serious, 'Megan. I know you moved my car. How did you do it? How did you move my car?'"

I stared at Sean, shocked. A spark of victory sputtered and poofed into flame in my soul. Mr. Carroll had been playing me! He'd been going all psychology on me. Ha ha! I gave Sean that high five.

As often happens in life, the real punishment didn't come from Mr. Carroll at all. Megan hadn't been one bit entertained by Mr. Carroll's reaction. When she'd said, "I didn't move your car. I don't know what you're talking about," our teacher had insisted that she did know and was lying about it.

"Then you come in," she said to me, "and he's all quiet and nice and, 'Amy, did you move my car?' And when you said you'd done it, he looked surprised, as though you're innocent and would never do anything!" That offended her. That offended her so much she told her father who - remember - was the chief of police. The chief of police was so offended that Mr. Carroll had called his daughter a liar that he threatened to phone Mr. Carroll and give him a less reasonable part of his mind.

No no no!

Thank God for Sean. Sean calmed down his Irish father, and Mr. Carroll didn't get clobbered over the phone.

And Mr. Carroll didn't even give me J.U.G.! Megan had a right to be annoyed.[1]

Mr. Carroll taught me something in those days, though, something I don't think he appreciated. I'd felt a blow when I thought I'd damaged our friendship. The fact that he'd (mostly) been messing with me didn't matter. I didn't want some stupid joke to ever hurt anybody for real. I didn't want to use my creative spirit to dream up stunts that pained those I cared about or made their lives harder or made them feel like the heel, the butt of everybody else's joke. I might feed Rocky Mountain oysters to good natured NASA guys,

---

1 For the record, Megan was an outstanding and faithful friend with whom I enjoyed many adventures, and she endured much excess drama in her life on my account.

but otherwise I'd be careful.

After moving Mr. Carroll's car that day, I made a decision. I decided that if I were filled with aggravation and cabin fever and needed an outlet for my pent-up creative needs, I wouldn't plan evil against anybody. Even entertaining, lighthearted evil. I'd plan good things. I'd plan things that made people feel valued. Even my retaliation against Mr. Carroll my senior year involved *cleaning* the O.M. room - not causing destruction.

Years after all these things, I faced a semester filled with Physical Chemistry, Organic Chemistry, Instrumental Chemistry, Statistics and Physics. I was in for it. I had Grand Canyon planning to do, but I desperately needed a creative outlet or I might spontaneously combust in the hallway one day, I didn't know. And the most convenient valve for releasing creative pressure was my favorite astrophysicist in the world, Dr. Zenith.

By the spring semester of 2011, Dr. Zenith and I were... okay. We were not best buddies, but we'd chopped out a relationship of mutual respect. It had taken time and patience and stellar work to free me from my position as the jawless hag fish of the classroom. I had risen from the rank of bottom-dweller to a place where I was permitted to speak before I was spoken to. In the world of Dr. Zenith, that represented great victory.

I had no desire to harm Dr. Zenith. I wanted to give him something to enjoy. There was no vengeance in my heart toward the astrophysicist, not like that day I'd cleaned the high school basement. Dr. Zenith had demonstrated a dedication to teaching students, to showing them how to *love* physics. The tension between us, though, I had long wanted to beat that. That was the true enemy.

Even back when he'd disdained me, I'd wanted to give Dr. Zenith candy bars. Between cabin fever and a chemistry-heavy schedule, and the fact that I truly appreciated the astrophysicist, a plot began to form in my heart. It started small, but eventually it climaxed in Matthew Caerphilly's pelting Dr. Zenith with Nerf bullets on the floor of his own lab.

CHAPTER 9

PLOTTING

AJ: Hey! Do you have any propane?
Kyra: Cocaine?
AJ: No, propane. I need propane.
Kyra: You and me both, honey.

"When the heck does Dr. Zenith turn 40?" I asked Matthew one day. "He's been 39 for a long time. He's going to turn 40 soon."

"I don't know," said Matthew. "How do we find out?"

"Mmm... I haven't been able to hunt it down through my normal cyberstalking sources," I chewed a piece of dried skin on my lower lip. Chapped lips, the bane of winter.

I turned to Kate, who didn't resemble the Apocalyptic Pale Horse of Death no matter what Dr. Zenith said. She had chopped off her waist-length hair over the winter and now sported a super jazzy pixie haircut. "Kate, how do we find out Dr. Zenith's birthday?"

"I don't know. But we have to."

We had to.

We finally did sleuth up the date of the astrophysicist's birth. We did not pickpocket him and check out his driver's license. We did not ask his mommy. We didn't break into any secret files anywhere. We found out his birthday, though, through means that were fair and square and legal and mostly above board. And I shan't tell you how. That's like... that's like a magician telling his secrets. We sleuthed up the date of Dr. Zenith's birth and set to work, Matthew, Kate and I. We had just one month, because Dr. Zenith was indeed turning 40, and his birthday fell conveniently, ohhh so conveniently, on the Friday before Spring Break. THE FRIDAY BEFORE SPRING BREAK! We could not have made that happen. We could not have conjured up such brilliance.

"I have Star Wars action figures we could hang from his door," said Matthew.

"Oh, that's a good idea. And we should… we should load his door with Three Musketeers bars." The man loved nougat.

"Doesn't he like Milky Way too?" Matthew asked.

"I think so," said Kate. "I don't know."

"He doesn't like Snickers," I said. "He likes creamy."

"Well, we should get both Three Musketeers and Milky Ways."

"Good."

"What else. What else."

"Well, you know how he said in class, 'I need a thwacking stick?' And somebody said, 'A whacking stick?' and he said, 'No, a *thwacking* stick.' He needs a thwacking stick."

"Oh yeah! A big thick one! And we should burn the letters into it, 'THWACKING STICK.' I have a wood burning kit."

"That's excellent. There's a lot of little trees on my property we could choose from. Next time you come up we'll pick one out."

"I love it. A thwacking stick. That's so great."

"Well, what are we … how are we going to give it to him? Are we just going to lean it up against his door with a big bow on it?"

"No, that's lame."

"Yeah, that's lame. We need something better than that."

"I know. We'll do a treasure hunt! The first clue can start at his door, and then he'll have to follow the clues."

"And there have to be enemies."

"Yeah, enemies! Because he's getting a thwacking stick! It's his weapon to fight off the enemies."

"This is going to be so much fun!"

We were like Peter Pan's Lost Boys. We were chemists and engineers-to-be, and we were planning a treasure hunt to give the astrophysicist a thwacking stick, Star Wars figures, and Three Musketeers bars. Throwing Dr. Zenith a 40th birthday party was completely necessary to make that excessively left-brained semester bearable.

In February, I began to write clues, clues that led to a variety of locations in the science labs and offices that Dr. Zenith could access.

-"Bones" - the skeleton in Flash's lab.
-The phases of the Moon above Dr. Zenith's lab door.
-The stuffed spherical cow at the back of his lab.
-*The Far Side* cartoon on the corkboard outside his office door.

One day in his office, Flash handed me a well-aged hardback book: *School Physics* by Elroy M. Avery.

"Check that out," Flash said. "It's a science book from 1895 that Dr. Stillwell gave me. There's an experiment in there that uses gunshots to test the speed of sound. Can you imagine trying to use that experiment in a lab today?"

I laughed as I skimmed down the offending page. "That's great. You'd get fired, but I think students would remember the lesson."

I silently noted the page number. Boom. A clue had to go into that book.

I scanned the young professor's office for another hiding place. Flash's room felt white and empty compared to the whirling buzz of Dr. Zenith's office next-door. Flash had only been teaching there two years; he needed more life in his barren cube.

"What is that behind you?" I pointed at the top shelf of his computer desk. "Is that a Starship Enterprise?"

Flash turned in his chair and gazed at the eight-inch-wide model. "Oh yeah. Yeah, Dr. Zenith gave that to me."

"Really? Dr. Zenith gave you a Starship Enterprise? I'd treasure that thing."

And the USS Enterprise. We'd have to hide a clue there too.

I wrote clues when I should have been doing school work.

> Jeffrey Gurden's goofing off
> It's Friday after all.
> I'll bet he'd use his holodeck
> if it weren't so darn small.

> Dr. Gurden's still too bored
> He wants to test the speed of sound
> Ask him for the lab he'd use
> If he had a gun around.

Matthew drove out to my property and we sacrificed a small tree for the Thwacking Stick. It was going to be a nice, big, brutal weapon. A veritable staff.

We then had to figure out where the clues led in the end. Where would we place the brutal cudgel? Dr. Zenith's lab? Dr. Gurden's lab? The stuffed Lorax in Dr. Stillwell's office? Ahah! Dr. Zenith had given Dr. Stillwell that Lorax! The Lorax would hold the Thwacking Stick.[1]

Clue List:
1) His door (spherical cow clue) - I
2) Spherical cow (center of the universe clue) - S
3) Center of Universe model (old physics book clue) - P
4) Old physics book (mean astrophysicists clue - crossword puzzle) - E
5) Mean astrophysicists (Star Trek toy clue) - A
6) Star Trek toy (phases of the Moon clue) - K
7) Phases of the Moon (Bones clue) - 4
8) "Bones" Skeleton in Dr. G's lab (CD for license plate)
9) Dr. Zenith's license plate ("THE TREES")
10) Lorax - (Thwacking Stick)

I penned one extra letter on each clue, nice and bold. If Dr. Zenith put all the clues together in order, they would say:

---

1 I recognize the sick irony, since we'd sacrificed a tree to make it – and the Lorax speaks for the trees. But, you know, the whole thing was twisted from the start.

I S P E A K 4 THE TREES.

That final clue would lead him to the Lorax.

Okay. All that was settled. He would win a Thwacking Stick for his birthday.

Sweet.

But there needed to be more than that. Dr. Zenith had to have a reason to *use* the Thwacking Stick. There had to be enemies.

"You know," I told Matthew and Kate. "Dr. Zenith says there's a prophecy he'll be killed by ninjas."

We'd have to get some ninjas.

And! And if ninjas were going to end Dr. Zenith's life one day, he needed a chance to beat one to death. We bought a Spiderman bust piñata and painted it black, and Matthew's sister Megan made katanas to attach to its back. I hunted through faces online and printed out a pair of eyes for the piñata. We then filled the ninja piñata with Dr. Zenith's favorite candy bars.

Phew. Between the Grand Canyon and Dr. Zenith's fourth decade outside the womb, I was saved from exploding cows.

Chapter 10

Scrapes and Sprays

Jared: My face feels like leather.
AJ: It … it is leather.
Jared: No, leather is the product. My face is the raw materials.

The children and I no longer lived in the van, by the way. We had a new kitchen, new hardwood oak floors, new windows around the house. I built a giant pine bookshelf for the kids' bedroom and painted the shelves bright blue and white. Our cabin was becoming a pleasant home.

My eight-year-old Sammy crashed his bike into the bushes the next weekend during a February thaw. He ran into the house and showed me his scrapes. I was happy he'd been outside crashing his bike, being an adventurous boy. I wanted him to feel brave and strong.

"Wow! That's awesome!" I approved.

My son isn't made of leather. He came inside a bit later and said, "Savannah would have had compassion on me."

I gave Samuel a hug. I asked him, "Would you rather I had compassion, or would you rather I said, 'You got scrapes from being a boy and crashing your bike. You're awesome!'"

He smiled. "I guess both."

I gave him another hug.

I'm leathery, but compassion is a good thing. I think it's good for me to tell my son, "Yayy! You're having adventures!" It's also good to treat him with tenderness and wash and bandage his scrapes.

Other people understand this more naturally than I do. I know it sounds like a normal thing, but I have issues. When I was in the 4th grade, little Susannah fell on the pavement while jumping rope and skinned her knees. She sat down and cried, and I watched her, genuinely puzzled. I didn't despise her or anything, I just didn't understand why she was crying. I skinned my knees all the time, but I'd jump up and run down to Mr. Face's office to get bandaged, then return to playing soccer on the pavement. As I studied Susannah in her distress, I said to myself, "Oh. She's in the 2nd grade. She's seven." I gave her that; she was a little kid, *that's* why she was crying.

I'm not sure what it is. I feel physical pain. I feel it, and I don't like it, but maybe it doesn't bother me as much as it bothers most people? I don't know. So, I have issues.

That same week, Dr. Vallo worked to help me get my nose washed out when I had a little organic chemistry accident. I was a competent adult and not an eight-year-old boy, but Dr. Vallo would have helped patch me up if I'd crashed my bike.

Dr. Vallo led us in an organic chemistry experiment that involved salicylic acid. As I scooped some of the fine powder from a brown glass jar, I spilled a bit and it puffed into the air. I was in the middle of inhaling as this happened, and I breathed in salicylic acid dust. Concerned, I scanned the bottle label to find out the dangers of concentrated salicylic acid and read a list of (frankly scary) potential side-effects.

Did I go to Dr. Vallo for help? No, that didn't even occur to me. I simply walked to the sink to wash powder residue out of my nose, worried about having the stuff in my lungs. Dr. Vallo didn't leave me to my self-sufficiency, though. He saw me splashing water in my face and kindly walked over.

"No. Use the sprayer over here." He led me to the eye-washing station.

"It's not in my eyes," I explained. "I breathed it in."

"It's okay. Go ahead and use it to wash out your nose."

Dr. Vallo didn't abandon me. He didn't hand me the sprayer and expect me to figure it out myself. He pulled out the hose to show me how to use it, and he - ha ha ha - he accidentally sprayed himself full in the face.

Dr. Vallo sputtered and laughed as water dripped off his nose and short gray beard. I smiled in appreciation as he handed me the sprayer and went for some paper towels. Then I washed out my nose.

The next week when I selected a compound for an experiment, my friend Katz teased me, "Now, keep that stuff *in the jar*." A rumor went around that I'd gotten chemicals in my eyes. I corrected a few people, but I finally let it go.

Thank you for your compassion, Dr. Vallo. I will work harder to be more like you.

I have another silly story that I want to add here just for the fun of it.

Dr. Vallo paid me well to clean his house and help his wife oil their hardwood floors and suck up their stink bugs, and I liked him as a person. He was a good sort of fellow. However, I honestly struggled to follow the material he taught in his Organic Chemistry class. He spoke English fluently, and I could hardly hear his Hungarian accent, but it was almost as though he used a different form of English than I did. I had to deliberately focus to understand what he meant as he chattered about benzene rings and carboxyl groups. I learned a ton during study sessions, because he had us write out organic chemistry problems on the board. Hands-on practice helped me tremendously. However, in class I struggled to understand the meaning of words as they flowed from his mouth.

One particular day that semester, Dr. Vallo singled me out to answer a question. To my dismay, he turned to me in class and said, "Tell me, why do I say, '***an*** aniline' or '***a*** phenol?'"

He had flashed a spotlight on me, and I felt exposed and vulnerable. I had been listening to the lecture, but I had no idea what he meant by that question. There was no context for it, and tension filled me as I scrambled in my brain to figure out exactly what he was asking. He had stressed the articles in front of the words though, so I said the only thing I could think of in that moment.

"You said, '*an* aniline' because 'aniline' starts with a vowel, and you said, '*a* phenol' because 'phenol' starts with a consonant."

And the class shattered into laughter. It was obviously a ridiculous thing to say, but I couldn't think of anything else!

Dr. Vallo stared at me, a blank expression on his face. In a gesture of hopelessness, he tossed his white board marker straight into the air. Ben Hackett shook silently in the seat next to me, his face going red.

"That might be the correct answer grammatically," Dr. Vallo announced. "But this is organic chemistry."

During that noisy moment, I worked over Dr. Vallo's question, trying to decide what he really wanted. Why did he say, "an aniline?" What was he asking? Then, woosh! I realized the simple point he was trying to make and quickly recovered. I said, "It's because there are many different *kinds* of anilines and there are many different *kinds* of phenols."

The gentleman's face relaxed, and he nodded and continued his lecture. I hoped my fellow students thought I'd merely made a joke and didn't realize I'd been accessing my internal resources of moron.

I'm sorry. This story still makes me laugh years later as I think of Dr. Vallo standing there, deadpan, tossing his marker into the air.

I had fun with all my professors. I had Grand Canyon planning, and Dr. Zenith's birthday planning. I was having a good time. But the third remarkable set of events that semester had a lot to do with Dr. Stillwell and a steady series of sessions on when to turn right and when to turn left.

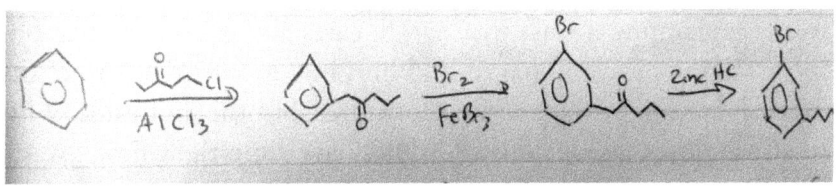

Figure 4: An example of a Friedel-Crafts Acylation and Clemmensen Reduction, organic chemistry at its finest.

Chapter 11

Turning Right

Journal entry, February 2011:

... There's a new physics teacher a-coming who gets to give a talk this afternoon, so I think it would be fun to go to that and the kids might like it too...
I hope she's nice.
I hope her eyes sparkle.
I hope her lab is made of gum drops.

I miss those days of birthday party plotting. They were warm, merry days. Lovely, cheerful days, despite the ice that still plagued the roads between our schools and the cabin. I stopped into Dr. Stillwell's office a couple times a week to work out the details of our class trip. Dr. Zenith and I were almost not-enemies. The world was a beautiful place.

Dr. Stillwell liked having me around, but he was a busy man. He taught a full schedule of classes, and he had a family and a dozen cats at home. He didn't enjoy much spare time, so I tried to leave him alone unless it was important to talk to him. I spent days avoiding him just to make sure I wasn't getting underfoot, even if we *were* planning a class trip. Dr. S. cares about his students, but to this day it's exceptional to find him in a moment when he isn't harried. He might want to stop for coffee, but the multitude of his responsibilities take priority. "I don't have time right now," was a constant theme for him.

Of course, Dr. Stillwell had swiped all the Lego warrior figures from around his Power Point station. Really! He took every single one of those Lego guys and hid them away. What's more, he threatened to chop off their legs and arms and go all Dr. Frankenstein on their little yellow bodies!

"Maybe I'll glue the legs back onto their heads or out of their backs," he said wickedly. (I knew it! He was totally jealous about Dr. Gurden's ghost!)

What started to amaze me that spring 2011 semester was that my knowing-in-my-knower got activated regarding Dr. Stillwell. I didn't know anything about him that wasn't my business. I didn't know whether his wife was happy with him or whether he was holding dinner parties. I didn't know if his truck was broken or if his appendix hurt him.

I seemed to be given a small window of information about him, however, as it pertained to *me*. I began to know when he wanted me to stop by his office. I could sense in my spirit when it was a bad time to visit him and when it was a good time. This was odd, because it was consistent, and I wasn't asking for it.

It started the afternoon that Dr. Sytil came onto campus to do her audition for the new physics professor position. I wanted to go watch her tryout lesson on magnetism, but I also wanted to avoid Dr. Stillwell. I'd decided earlier that day that I'd purposely stay out of his way and leave him alone. I didn't want him getting sick of me.

Matthew had kindly grabbed the kids from school, and the three children joined me, ready to sit on tall stools and take in a college physics lesson. I debated about it inside myself, though. Maybe we should leave. The kids and I walked down the hallway, but I wasn't sure whether we would take door number one into the parking lot or door number two into Flash's physics lab. Walking down the hallway, though, my mind got made up for me. A sort of confidence washed through my chest.

I realized, "Dr. Stillwell *wants* to see me today."

You don't understand. Sometimes he was glad to see me, and sometimes he was irritated and didn't want to see anybody! I never knew whether I was going to get Dr. Jekyll or Mr. Hyde with him, and it's still an issue as I edit this for print.

That day, I got hit with an understanding as I walked along. I had no physical reason to think it, but I knew Dr. Stillwell wanted to see me. "Well, then okay," I thought. "I won't avoid him."

The kids and I continued into Flash's lab and watched Dr. Sytil's presentation. The cheerful woman did a great job with her lesson, even taking time to involve my children by giving them magnets to play with while she taught. Of course, you want to know if she splashed me with life, and I apologize, but I have to refrain from answering that question about Dr. Sytil.

We all hung around after the lesson to chat, and during this time my son Sam pulled me down and whispered in my ear.

"Can I get my Legos back from Dr. Stillwell?"

"It's okay. You can go up to him and ask him yourself. He'll give them to you."

Little Samuel tried. My eight-year-old eased up beside the geology professor, but he felt shy and struggled to speak his request. I finally relented and went to his rescue. I approached the good doctor and said, "Excuse me, sir, but we have to talk about an exchange of prisoners."

Dr. Stillwell tipped his head back and laughed out loud. "Is *that* what he was trying to ask me?"

I nodded.

"Okay," he said cheerily. "We'll go get them."

I gathered my little troop, and the kids ran down the hall toward his geology lab.

"Where have you *been*?" Dr. Stillwell asked as we walked along. "I never see you anymore unless you're rushing off somewhere."

"You know, it's normal - when students aren't in your classes - not to see them," I reminded the good doctor.

After that day, the day of Dr. Sytil's lesson, Dr. Stillwell finally opened up about wanting me around. He plain let himself be free about it, and it was the best thing about that spring semester.

"Dr. Gurden, will you please excuse me?" he'd say when I

appeared in the hall, "I need to harass this student."

These little things made me laugh. Dr. Stillwell might be talking to somebody in the stairwell as I trotted down from the second or third floor. When he saw me, he'd immediately end their conversation and step down the stairs with me. Or, there'd be another student in his office when I'd walk by, and he'd say, "Amy Joy. Come here!" He'd tell the other student, "Excuse me for a second. Can you sign that out in the hall? I need to talk to this young lady." One time I passed him in the hallway, and he grabbed my elbow and turned me around so I'd have to walk with him.

"Why are you turning me around?" I asked him.

"It seemed like a good thing to do," he answered.

Hah!

Normally, you'd think "Ooh. Creepy old man." No. No, it wasn't creepy. We'd become friends. I thought back to that dream I'd had and the closeness and warmth and innocence of it. It was innocent. It was a fun, warm friendship, one of those rare treasures in life.

And it was odd for two reasons. First, Dr. Stillwell despised people who believed in a real, personal God. He despised people who believed Jesus was coming back one day. He despised everything I stood for, and he knew that mine was not a nice, respectable Sunday pew faith. I claimed to have successfully commanded a fire to calm in Jesus' name. I believed God still did things in people's lives! I would have expected Dr. Stillwell to slam his door in my face, and yet he welcomed me.

And second, the Holy Spirit seemed to be encouraging the whole thing. How much fun was that!

Chapter 12

Spiritual Empathy

It was Randy who first encouraged me that God talked to me, long before I ever considered marrying him - this odd, bearded man from West Virginia.

Randy and I were just friends in early 1999. He drove a rusted-out white van that we called the "Spotted Dog." It had brown carpet in the back instead of seats, and every Friday night we drove around the county collecting small hordes of teenagers for youth group. Hordes. Like Mongols.

"Amy Joy and Randy sitting in a tree…K-I-S-S-I-"

"No," I informed the foolish 15-year-olds. There was no way. I was not remotely interested in Randy.

Our woman's pastor Patsy told me, "Amy Joy, Randy is handsome."

"No he's not, Pat."

"He *is*," she said. "In two years when you're walking down the aisle-"[1]

"No, Pat."

Randy and I were friends, and it was wonderful. I liked hanging out with Randy. There was no stress, no worries about whether he liked me or I liked him. We had a perfectly lovely, relaxed relationship, and dragging romance into it would have ruined everything. Randy was easy and funny, and I liked it that God talked to him. God didn't talk to me. Randy had visions. I didn't have visions. Since I couldn't hear God, it was nice to hang out with somebody who could.

"God doesn't talk to me," I said to Randy one day.

---

1 Pat was wrong. We got married the next year.

Spiritual Empathy

"Sure He does," he said.

No. Randy didn't understand how unique he was.

"God talks to me through the Bible," I explained. "That's what He already told me. He's not going to talk to me like He talks to you. Instead, He gives me verses when He wants to say things to me. He gave me the verse Jeremiah 31:3: '*I have loved you with an everlasting love. I have drawn you with cords of lovingkindness.*' He loves me, He just doesn't talk to me except through the Bible." My Heavenly Father had made it clear that He simply wanted me to trust Him.

Randy didn't appreciate how hard I'd tried to get God to talk to me. He didn't realize that it was a sore spot in my heart, and I'd resigned myself to disappointment. I didn't even expect God to talk to me anymore. I didn't even hope. God talked to Randy, so of course he thought nothing of saying, "Sure He does," as though everybody heard from God like he did.

I don't speak in tongues. I don't have visions. I've never prayed for somebody and watched them instantly healed before me. Randy had those gifts. 1 Corinthians 12-14 explains that there are a wide variety of spiritual gifts, and every believer's Holy Spirit superpower is an important piece of the whole Body of Christ. Some of us are eyes and some of us are toes or spleens or muscles. Not one of us should think we're insignificant, because we each have an important job to do as part of the whole Body. I didn't hear God the way Randy did, but God had given me gifts too. I just had to learn about them.

People may not appreciate the *variety* of spiritual gifts out there to be had. Isaiah and David both had prophetic gifts, but David also had courageous faith and musical ability and talent in leadership. In Exodus we learn that Bezaleel and Aholiab were given divine wisdom as artists.[2] Aholiab was an engraver and embroiderer, and that was a divine calling too, because he had the great and unique task of fashioning the Tabernacle, the tent that housed the Ark of the Covenant.

My brothers Baron and Shadow are excellent artists. Shadow links his skills back to his childhood, when he said in his heart, "I'm

---
2   Exodus 36:1; 38:23

going to draw this well because God wants it to be great." Shadow has found that when he depends on his own awesome skills, his artwork turns out awful. When he paints with the attitude that his creative abilities were God's idea in the first place, he produces some amazing work, like this painting of the U.S.S. New Jersey.

Figure 5: "New Jersey En Route To Qui Nhon." Our father commissioned Shadow to paint it, and it takes up one wall in Dad's living room.

In the 1981 movie *Chariots of Fire*, Olympic runner Eric Liddell has a sister who wants him to be a missionary, and she's excited that he's going to China. When he explains he'll be running more first, she's not pleased that he's training for the Olympics. Eric tells her, "Janet. Janet, you've got to understand. I believe God made me for a purpose. For China. But, He also made me *fast*, and when I run, I feel His pleasure."

There is a world of spiritual gifts and blessings out there. Those gifts come in a wide variety of forms, and I'm convinced that every one of us has something special, something precious that God has made for us to do. We go around feeling insignificant, and we don't realize how important we are to the people around us. God didn't work in me the same way He worked in Randy, but He had other things for me to do.

Randy told me, "Sure He does," and soon I was given a new lesson on my own spiritual gifting, one that came in a thoroughly unexpected form.

A couple of days later, I was hit by a strange emotion. There was no outward reason for it, no natural, obvious cause. Just minding my own business, I was overwhelmed by a horrible insecurity. It swamped over me like a soggy beach towel, and I tried to fight it off, but I couldn't shake the alien sense of being no-good. Remember, I've got what Dr. Stillwell calls "unbridled self-confidence." It doesn't matter how unimportant I have felt to other people, I've never suffered from a confidence problem. I like myself just fine. Yet, here I stood under a blanket of worthlessness that took away my breath and held me down.

I fought it for two days, tried to pray, used every manner of self-encouragement I knew. Nothing worked. It held on like an enormous tick with its head buried deep.

I wondered, "What on earth! Where is this coming from?"

On Friday, I walked into an auditorium where my friend Jenny Hunt sat on the stage. I saw the side of Jenny's head clear across the room, but I didn't see her face or her expressions or hear her talking about anything. I'd run into a wide variety of people during my two days of insecurity, but as I walked into that auditorium, I had a revelation.

"It's Jenny! It's Jenny who has been feeling this way!" I knew it in my knower. I had no doubt that Jenny had been feeling worthless, and I'd just experienced a two-day taste of it so I could understand what she was going through.

I strode to her and said with some excitement, "You've been feeling really horrible about yourself lately, haven't you?" (I wasn't so good at easing into things.) I tried to detail the precise emotions that she had been facing, carefully hunting for words that would exactly describe the struggle inside her. She hardly even spoke. She just nodded, confirming from her grieved heart that it was exactly how she'd been feeling.

I would never have thought to ask Jenny if she felt bad about herself. It wouldn't have occurred to me. I'm leathery. I hardly pay appropriate attention to my own pain, let alone suspect the deepest struggles in other people. I didn't even offer her any comfort when I talked to her, I just said, "Wow. You've been feeling this way, haven't you!"

After that, the foreign emotion evaporated. Foof! Gone. I suspect that's what Jenny needed most, though. She needed to know she wasn't alone, and that God was paying attention to her.

A month later it happened again. This time, I struggled with arrogance. I don't normally have problems with ego, but this go-around I couldn't shake a sense of superiority over other people. I kept looking down on them and despising them. It was awful. I hated it, but I couldn't get it to go away. Again I thought, "What on earth! Where is this coming from?"

Once more, I spent two days struggling against the unwanted, foreign emotion. Once more, I saw another friend clear across the room and said, "Ah! It's *that* friend who is struggling!" So, I went up to the friend and described her feelings based on my own two-day experience. Once more, I was right on, and once more the negative emotion disappeared and caused me no more trouble.

The third time it happened, it didn't take me two days.

We'd gone camping with the youth group outside Eugene, Oregon, where we slept in big open sheds and drank water from a pump and swam in the river. At night, our group of campers gathered on wooden risers out under cedars that rose high overhead in the dark woods. One night during worship, I stood on the risers and enjoyed singing to God with all my heart. I had a rare sense of peace and freedom in my worship, as though the spiritual sky above me had opened clear to Heaven without the normal dark clouds that blocked my view. We were simply enjoying each other's company, my Heavenly Father and I, and I rested in those precious moments of closeness and comfort.

Then, in an instant, my peace vaporized with an invasion of

doubt. In one second, my freedom and joy were shattered, and holes burst through my faith. I feared that God wouldn't answer any of my prayers.

I thought, "What on earth! Where is this coming from? I haven't felt like this in years…" I scanned over that crowd of thirty people below me and spied Courtney Gorman. "Ohhh," I realized. "It's Courtney." It didn't take two days this time. It took all of twenty seconds. I stepped off the risers and prayed for her.

How did I know it was Courtney? Nothing specific. Nothing about her appearance. I just knew it was her. I saw her and had confidence she was the one struggling with an invasion of doubt.

I felt like I'd gone through a training session. God cared about these dear ladies, and He let me join in on His desire to touch them and heal them. I don't think God *gave* me those feelings. I think He simply opened me up to the other person – a sort of empathy super-connection. He wanted them to know He was there and willing to help them. And He let me be a part of it. How great was that?

"See?" Randy said when I told him about it. "See? I told you God talks to you."

Over the years, I've grown sensitive to the Spirit of God. I've practiced listening to Him until it's become habit.

One good example is when my children vanish. Every year at the fair, I lose at least one of them. Zeke dashes off without asking, or Samuel doesn't meet me where he's supposed to. Let's say I drop my little Savvy's hand for a moment, and she dashes off. I go to grab her hand again, and she's gone. I spin around, scanning the crowds for her. I pace down the grass between the carnival tents, hunting through the faces at the fish bowl ping pong game, searching up and down the line at the monster slide. I whistle my distinct, loud whistle, the ones my kids recognize. Nothing. Did she run into the fun house? Did somebody snatch her? In my mother's mind, my six-year-old daughter is already stuffed into a van with duct tape on her mouth. That's my fear - that's every parent's fear when a child goes missing in a public place full of people. The panic overwhelms me as I look over the crowds and think, "Dear God, where is she?"

I've learned, of course, that panic doesn't help anything. I have to calm down and relax and trust God to show me where she is. I panic every single time, but the panic gets in the way. I have to trust God.

"Lord...where is she?" I listen in my spirit, quiet my heart and wait.

Then I'll know. I'll see it in my head where she is.

"Oh. She's at the carousel. Okay. Thank you!" So, I go to the carousel, and there she is. That's happened more times than I can remember.

In the spring of 2011, my knowing-in-my knower was activated regarding Dr. Stillwell. I started to know when it was a bad time to stop by his office because he was too busy, and I knew when he missed me and wanted me to show up.

CHAPTER 13

DEEPER DIMENSIONS

"If the wave-particle duality of light doesn't keep you awake, I don't know what will."

- Dr. Flash Gurden

This universe is a strange strange place. I'm not talking about the kinds of people who show up on Capitol Hill in Seattle. A grown man walking around in nothing but a large diaper? Disturbed, but not too hard to explain. I'm talking about what happens around us every second at the quantum level, where Newton's laws break down and nothing makes logical sense. The quantum world is a weird place.

I am absolutely convinced that the real world involves additional dimensions. More importantly, I believe we actually *exist* in additional dimensions. It's like we're living in water, but we can't see below the surface. A fifth or sixth dimension is difficult for us to conceptualize (the theoretical physicists say they are wrapped up tightly as superstrings, which doesn't help our imaginations much), but - bottom line - there is more to us than we can see with these eyes.

I've known things I have no physical way of knowing, and that requires an explanation. How is it that I can sense things beyond my physical fingertips? In a 1991 book,[1] writer Michael Talbot notably tried to explain his own supernatural experiences by arguing that we live in a holographic universe. That was his explanation. Talbot used an entire 338-page book to make his explanation, and I won't

---

1   Talbot, M. (1991, 2011) *The Holographic Universe*. New York: Harper Perennial.

go into that much detail. I do want to take a few pages and suggest that we live in a multi-dimensional universe, and we're blind to our full dimensionality. I believe there's a part of me that extends into the spiritual world, but I'm like Helen Keller. I can feel around, but I'm spiritually blind and can't see into those dimensions. I'm not completely deaf, but I'm definitely hard of hearing.

Extra dimensions are hard to grasp, but I think we can understand the idea of them a little if we just look at Garfield. You know, Garfield from the comics? The fat, orange cat that hates Mondays and loves lasagna?

Garfield lives in two dimensions on the flat page of a comic strip. I propose that if Garfield really were alive, he could only see his world in one dimension. He'd see a world of lines. *We* are able to see the full two-dimensions of Garfield, because we have the power of the third dimension to look *down* at Garfield. We can see the fullness of his orange fatness. We can even read his thoughts in his thought bubble.

But, Garfield doesn't enjoy three-dimensionality. He can only look forward and backward. He can only see Jon as a multi-color line. Odie's drool shoots at Garfield as one-dimensional dashes. We can look down and read his bubble-thoughts, but Garfield can't see what Jon is thinking, because he can't look past the bubble outline. Garfield has no clue he's missing out, and he can't appreciate the full two-dimensionality of himself because he's completely used to his one-dimensional world of lines. We who have greater dimensionality can see Garfield as he is, but he can't see himself or Jon or Odie as they are. See what I'm saying?

When God talks to me, to my spirit, I believe He's talking to the whole me, the me that extends into those additional dimensions that are beyond our eyesight. When angels come and go, I believe they're visiting from those dimensions.[2]

By the way, quantum physics suggests that there are many more than just the four dimensions we directly experience. What follows is a brief, mercifully brief, introduction to quantum mechanics,

---

2   See "Angels and Witnesses" in the Appendix.

because quantum mechanics gives us clues about the deeper reality of our world. Subatomic particles point beyond the whiny, wimpy four dimensions we see with our eyeballs. Besides, I know you've been wondering about Schrödinger's cat, that poor confused kitty, and I'm here to help. Seriously, who doesn't want to know a little bit about quantum theory?

THE DOUBLE-SLIT EXPERIMENT:

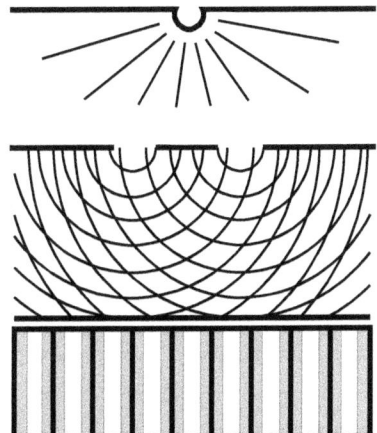

Figure 6: The interference of light waves creating a pattern of light and dark bands.

Consider the famous double-slit experiment, first performed by Thomas Young in 1801 to demonstrate that light is a wave. The variations on this experiment have shown us odd things, the sorts of things that make us question the very nature of reality.

Let's say we shoot a stream of photons at a plate with two slits, like in Figure 6. The photons create a wave pattern behind the plate - pew pew pew - because light behaves like a wave. As the photons pass through each slit and hit each other, they form an interference pattern of light and dark bands on the screen behind the plate.

Dark. White. Dark. White. More light. Less light. No light.

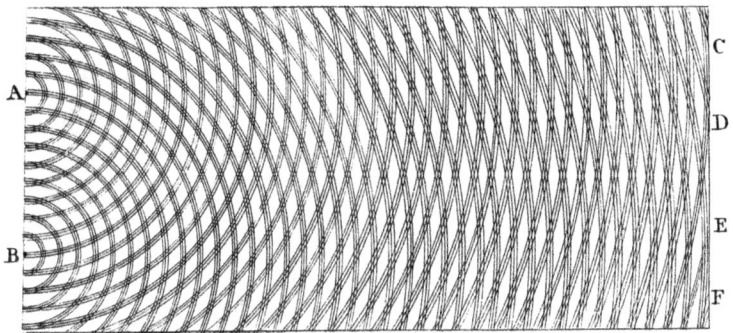

Figure 7: An 1803 sketch Thomas Young made of the interference pattern of two waves. The A and B represent the two light slits, and the C, D, E, and F line up at the dark bands.

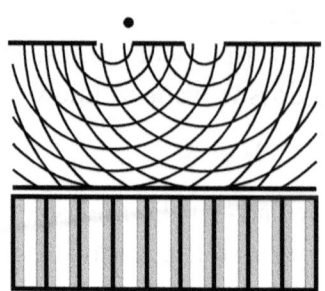
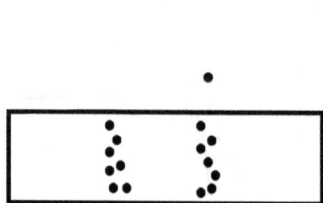

Figure 8: The interference of waves formed by single photons creating a pattern of light and dark bands.

Figure 9: Single photons creating a bullet pattern when monitored by a detector.

We see this phenomenon all the time with wave patterns, even with the waves made by throwing two pebbles into a pond. The same principle works for light waves. When a wave crest runs into a wave trough, they cancel each other out. That makes a dark band. On the other hand, two waves hitting each other combine to form a bigger wave together. That makes a light band. This is an interference pattern.

Here's where it starts to get weird, though. If we slow down our shooter to spit out one photon or electron at a time, each one hits the screen and makes a little light point like a bullet. Which is fine. Except, if we do this long enough, we find that the points of light add up to form an interference pattern just like the stream of photons!

What! That's nuts! How can a single particle interfere with itself? It knows both slits exist, but does it split and send a twin through each slit? How can individual particle points add up to a wave interference pattern! How is that possible! It's acting like both a wave and particle at the same time. That's crazy.

Fine. Be that way, little photon. We'll put detectors at the slits (or across the room, it doesn't matter) to keep track of each particle as it zips through. We'll *watch* the photons go through both slits at once!

As soon as we do this, though, the pips no longer form a wave pattern. The very act of looking at the photon appears to force it to make a choice. The wave function breaks down, and a bullet pattern

Figure 10: Schrödinger's cat in the box thought experiment. If we peek into the box, we observe that the cat is either alive or dead, but if we don't look, it's both alive and dead.

forms instead, with no wave pattern build-up.

Okay. Woah.

This experiment has been performed a multitude of times with the same puzzling results. Photons and electrons act like both particles and waves. Protons and neutrons also act like particles and waves. In fact, every one of us is both a particle and a wave (but the wavelength isn't easily detectable at our large size). In the world of subatomic particles, the wave nature becomes obvious, and the very act of watching particles affects the way they behave. So strange. It's like they're embarrassed to do their magic in front of us.

THE CAT

Famous Austrian physicist Erwin Schrödinger likened the situation to a cat in a box. He imagined a box with a cat inside. In the box with the cat is a bottle of poison, a lump of radioactive material, and a detector. If the detector reads that a single atom has decayed, it causes the bottle of poison to break, which kills the cat. If no atoms decay, the cat remains alive.

What's the point, Erwin Schrödinger (you cat hater)? In his thought experiment, there's a 50/50 chance the atom has decayed. If we open the box to look at the cat, we see that the cat is either alive or dead. One or the other. However, as long as we're not looking, the cat is both alive *and* dead, a state called "superposition." What?! It can't be both! Yet, it is. All subatomic particles are both waves and particles until we look at them. As soon as we look, the wave function breaks down. What does that mean for the nature of reality?

It means that Dr. Zenith can wear a t-shirt that says "Schrödinger's cat is dead" on the front and "Schrödinger's cat is alive" on the back, and the nerds around him snicker.

"You know what I think that means?" my brother Baron chimes in. "That we really are God's television show, and we only exist because He's watching us. Put that in your book and smoke it."

Thanks, Baron. Thanks.

Electrons aren't so much waves or particles as they are electrons doing their thing, and we don't have the visualization language to explain them. We can write pretty mathematical formulas, but not paint pretty paintings.

I honestly think the fundamental problem is that we can't see in enough dimensions. I've long pictured subatomic particles as the pointy ends of strings on a rug's fringe, all connected to each other down where they reach the rug itself. They look like points to us, because we can't see the rest of the rug. If we wiggle the carpet, the energy makes a wave and ripples along to those stringy ends. In real time, quantum particles act like both waves and particles, but we see them from one angle or the other, either as a vibrating stringy wave or as a string-tip particle. I'm sure the theoretical physicists could make adjustments to my analogy, but that's the best I know to understand what's going on. Anyway, that's enough for now.

Fear not, I'll come back to it.

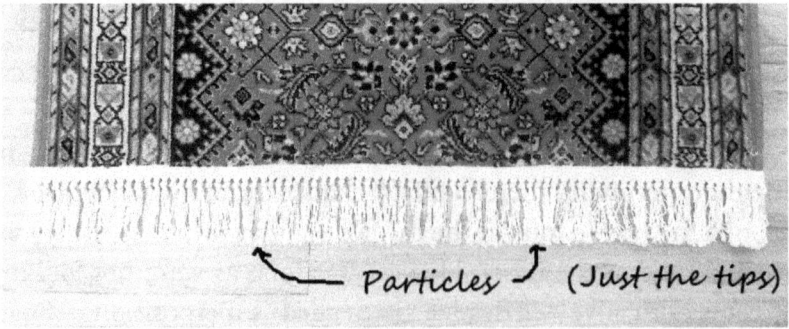

Figure 11: The simple analogy of the throw rug if atomic particles are like strings.

Chapter 14

Plotting

Our birthday "project," as Matthew called it, was so much fun. Dr. Stillwell initially rolled his eyes when I told him about the party we were organizing. "Why do you roll your eyes when I plan these things?" I asked him after a bit. "Don't you like them?"

"Oh, I love them…" He had some misgivings he didn't vocalize. Maybe he thought I was too busy as it was, that I should have been doing my schoolwork instead of putting together complex festivities. Dr. Stillwell didn't understand. Giving Dr. Zenith a fun 40th birthday made me a better, calmer member of humanity.

And of course we had to pull in other professors to enjoy Dr. Zenith's big day with us. Dr. Stillwell agreed to do his part and put the Thwacking Stick on his stuffed Lorax the morning of the party. Flash agreed to let us hold the party in his lab. Flash also had the crucial job of talking to Dr. Zenith at noon, keeping him occupied while we finished readying the food and balloons, gifts and cards until the ninjas arrived and attacked.

The prophecy that Dr. Zenith would be killed by ninjas would not come true on his 40th birthday, but he definitely needed a practice round. We had five ninjas scheduled to attack him in his own lab, and when he chased them, they'd lead him back to Dr. Gurden's lab where everybody had gathered for the party.

It was going to be so great!

Kate ordered a massive cake. We bought decorations and candy and carefully planned out the night and morning before the party.

"I've talked to the janitor," Kate said. "She's going to let us into

Dr. Zenith's lab so we can place the clues that go there."

Excellent.

We filled the ninja piñata with Three Musketeers bars. Kate and Matthew invested in several large Nerf guns.

"Do ninjas use Nerf guns?" I asked.

Those ninjas did.

Of course, certain things did go wrong. Things always go wrong. Matthew took all the bark off the Thwacking Stick, removing half its powerful diameter, and it suddenly looked like a weapon against insolent children rather than the warrior's staff it was meant to be. I saw that stick, and disappointment overwhelmed me. It wouldn't do, it wouldn't do at all!

"We've got to get a new one," I told Matthew. "We'll have to find a bigger one that will still be thick and sturdy after you remove the bark."

"But it won't have time to dry!" he protested. "It will crack!"

It didn't matter. We were getting close to the big day, and if the Thwacking Stick had a few hairline cracks, so be it. We hiked through the woods on my property and sacrificed a stouter small tree.

One of my proudest moments during the whole project was our concealment of clues in Dr. Gurden's office. Kate and I could have handed Flash the clues and said, "Hey, could you tuck these into their hiding spots for us?" We could have done that, but then Flash wouldn't have enjoyed the surprise. I wanted Flash to feel completely natural and unsuspecting when Dr. Zenith came clue hunting. Which meant we'd have to plant the clues without his knowledge.

Two days before the great party, Kate and I stopped by Dr. Gurden's office. Kate asked questions of our young professor while I examined the equations on his white board.

"Hey," I said casually. "Can I see that old textbook again? The one that had the speed-of-sound experiment in it?"

Flash handed it over and continued to talk to Kate. I opened the book to the appropriate page and slid the cream-colored clue into place. I glanced over the experiment again, then closed the volume and handed it back to Flash.

He didn't even look at it as he tucked it back on his shelf. Sweet.

I continued my survey of his white, boring room. Flash had posted the poem we'd given him for Thanksgiving on the back of his door, so he could read it when his door was closed. That warmed my heart, but the man's office wanted more life, more color in there.

(Eleven months later, Dr. Stillwell cleverly "closed" Flash's door behind him as the two of them walked out. A minute later, Dr. S. let Kate and me into Flash's office and stood guard while we climbed onto the physicist's filing cabinets and desk to rim his ceiling with panoramic posters of Zion National Park and Bryce Canyon. We gave Flash a plant for the empty flowerpot his wife had made, and under its shade we placed a jar with a guppie named Kujo. Kate and I have little respect for other people's personal spaces on their birthdays.)

That 9th of March in 2011, I wandered past Flash and Kate to study the one bit of flash in his room, a poster of a flying saucer on his far wall. "I WANT TO BELIEVE!" it declared in white block letters. I picked up the Starship Enterprise model and dropped a folded note under it, then returned to the white board and perused its equations again.

"Thanks, Dr. Gurden," Kate finished up as we headed out.

"Sure thing," he said.

Success! Kate and I grinned at each other. We'd inoculated his office with such surreptitious cleverness! He was the vector to pass the clues onto Dr. Zenith. Haha, biology strikes again!

Of course, Matthew called me on Thursday, massively upset about the Thwacking Stick. "It cracked!" he wailed. "I had to set it near the wood stove so it would dry, but it cracked just like I said it would!"

I tried to reassure him. "It's okay, Matthew. I'll come over and look at it, but I'm sure it will be okay. A crack will give it character."

I inspected the staff that evening and thought it looked great. The "crack" was hardly noticeable, and Dr. Zenith could knock out some bad guys with that thing.

"Your handwriting is better than mine," Matthew groaned, still unhappy. "Would you burn the letters into it?"

Smoke filled the Caerphilly dining room as I set the staff on the table and carefully, so carefully, drove a sizzling metal tip through the dried hickory sapwood. Within a few minutes "Thwacking Stick" smoked at us in strong black letters. We drove a screw down into the tip-top of the staff, lacquered it, and hung it up to dry overnight. When we took it down the next day, we left the screw in. I hoped Dr. Zenith would value his brutal birthday weapon.

Chapter 15

The Zenith Party

Disasters struck Thursday, March 10th, one day before Dr. Zenith's party. One struck lightly for us on a personal level, and the next day others struck massively and brutally for the people of Japan in ways that had international repercussions.

It had been raining. It rained and rained and rained for days, and water backed up through the pipes and flooded Dr. Zenith's lab in its exposed location on the first floor. Kate and I walked past his door and spied all his computers sitting up on desks.

Oh no.

On a far more serious note, soon after midnight EST Friday morning, a 9.0 earthquake hit Japan, moving the main island of Honshu a full 2.4 meters to the east. The earthquake caused a tsunami that devastated large portions of the Japanese coast, killing nearly 16,000 people and displacing 230,000 others. The tsunami also indirectly caused the meltdown of three reactors in the Fukushima Daiichi Nuclear Power Plant. The disaster had long-lasting results, and radioactive water from the wounded power plant continued to leak into the Pacific Ocean for years, infecting life throughout the seas of the world.

In our own little bubble, the flooding of Dr. Zenith's lab meant all kinds of bad things. It caused stress for Dr. Zenith and threatened his electronics, and it also kicked our plans every direction. Dr. Zenith wouldn't be teaching in his own lab the next day. He'd be teaching in Flash's lab, which meant that we couldn't set up the party there. And Dr. Stillwell might not get back from Charleston in time for us to use his room.

Ahhh!

There was also the terrible possibility that Dr. Zenith wouldn't appreciate an elaborate party while chaos dominated his life. I imagined his saying, "Thanks for the thought, you guys, but I have too much on my mind right now." We'd been planning for nearly a month, but I saw a dreadful anticlimax coming on.

We were tenacious sorts, though, and didn't let pools of water spoil our plans. Thursday evening, the janitor unlocked the door to Dr. Zenith's damp lab long enough for us to place all the clues in their hiding spots at the spherical cow, the phases of the Moon poster above the door, and the Copernican model of the universe under his Power Point screen. We slid clues under the hat of the skeleton in Flash's lab and behind *The Far Side* cartoon on the corkboard outside Dr. Zenith's office.

Matthew hung a Yoda-themed "Happy Birthday" banner above Dr. Zenith's office door while Kate and I taped snack-sized Three Musketeers around his doorframe. As I stuck tape on the silvery bars of nougat-filled chocolate, I prayed heartily, "Lord, please make it fun for Dr. Zenith. Let it be a blessing for him and not simply a big irritation." I wanted the astrophysicist to feel loved for his 40th birthday, and as I prayed, confidence filled me that God wanted it too.

I had a big Organic Chemistry test that Friday and had to study all night, but I was at Kate's house before dawn the next morning. We sliced up tomatoes and red onions and peppers and arranged meat and cheese on trays. We had all the good things for build-your-own sub sandwiches, with Sun Chips and pop and cake. We'd taken Star Wars posters and attached them to poster board as giant birthday cards. We made everything ready, and then we zoomed to school.

I called Dr. Stillwell as I walked from my car, and he answered the phone. Hoorah, he had returned from Charleston! We could hide the Thwacking Stick on his Lorax as planned!

On the other hand, I'd left three of Dr. Zenith's clues at Matthew's house! They were the clues I'd had to wait until the morning to position, including the clue that went on Dr. Zenith's license plate and the note that went with the Lorax.

I knocked at the geologist's office door and rushed in with the Thwacking Stick. "Dr. Stillwell, could I borrow some paper?" I rested the staff on the yellow arms of the Dr. Seuss creature on his office back counter, and Dr. S. handed me several sheets without a word.

Okay. Okay. Rewrite the missing clues. Heart pumping. I inked out a Zenith-level dramatic victory note that warned the man of his enemies, and I nestled it under the Thwacking Stick:

You have proved your worth, warrior of intellect and cunning, as we had faith you would. Now take this staff, keep it near all day (if not longer). It will prove valuable if swung swiftly when the enemy attacks.

I jotted out the other two clues, thanked Dr. Stillwell earnestly, and bolted down the hall to the staircase. On the second floor, I stood sentry at the window until Dr. Zenith's car rolled into its usual spot below. Unaware of my excess supply of adrenaline, the astrophysicist climbed out calmly. Slowly. Maddeningly slowly. He collected his things, then ambled toward the school door. As soon as he disappeared inside, I quick-stepped down the back stairs, dashed out, and stuck clues on his windshield and license plate.

I didn't even walk down his hallway after that. I couldn't bear to see him face to face. Still, I wanted to know what was going on, so I raced up to the third floor and grabbed a new chemistry student. "Bryce! Can you run down to the first floor and tell me what's happening down there? Just walk down the hall and whatever is going on, let me know?"

I used a more demanding tone than necessary, but Bryce was still a brick and did just what I asked. He returned five minutes later and said, "Dr. Zenith is in front of his office door. He was on his cell phone, saying, 'They've found me out. I'm standing in front of my door and there are Star Wars figures hanging all over it.'"

I grinned in delight. Until that moment, a part of me still worried we'd gotten the date wrong.

Fifteen minutes later, Kate texted me.

"8:46: Dr. Zenith and his wife are both on the scavenger hunt. He appears to be having a blast with it... When I got there it was on the first clue in Dr. Gurden's office. And Dr. Gurden was shocked it was there. This is so awesome!!!"

I danced around in a circle.

Kate later shared, "He went in and asked Dr. Gurden, 'A small holodeck? Where would I find a small holodeck?' Dr. Gurden slowly turned in his chair and stared at his own Enterprise model. It was so great."

Dr. Gurden helped his colleague more than once, astonished to find clues in his own office. Victory!

Another clue bellowed, "Jim! I'm a doctor not a bricklayer!"

Dr. Zenith read it out loud to Flash, puzzled. "Dr. McCoy?"

"Oh," Flash reminded him. "Bones."

Dr. Zenith bee-lined for Dr. Gurden's lab, where he found the CD case we'd stuffed under the hat on the skeleton's skull.

In the meanwhile, I rushed around to the offices of various professors and warned them, "Dr. Zenith's lab is flooded. We had to move the party." I then jumped down the stairs to Flash's lab to settle into my seat for class.

At 9:05, five minutes before class started, Dr. Zenith marched in with a pile of clues. He stared straight at me and asked, "'I speak for the trees?'" His eyebrows were furrowed in puzzlement. "'I speak for the trees?'"

How did he know! How did he know I was involved? I knew he was bright, but why would he suspect *me* of all people?

Oh wait. Wait. His peripheral vision eyes! Darn, it's always hard to tell which direction he's actually looking with those things! He wasn't staring at me at all. He was talking to his buddy Dr. Gurden - on my left - and I happened to be in the path of his other eye.

Dr. Zenith laid out the clues on Flash's lab counter, and the two men hovered, pondering. Dr. Zenith had figured out that the individual dark letters on each clue spelled out a message.

"I speak for the trees..." Flash repeated. Then his eyes lit in recognition. "Oh, I know what it is."

"Well, are you going to tell him?" the physics students all asked.

"He's got Google. All right, let's get class started."

With that, Dr. Zenith marched out of the room.

Another five minutes passed, and the astrophysicist returned. The Thwacking Stick rested proudly over his shoulder. "Thank you," he told Dr. Gurden, staring at me with one eye. "Thank you very much."

Dr. Zenith had thanked Flash and not me, but I still took those words personally. I accepted the astrophysicist's gratitude in that moment, even though he had no idea that I'd burned "Thwacking Stick" into the staff he carried.

That didn't end it, of course. Dr. Zenith had received the Thwacking Stick for a reason; there were enemies about. The ninjas had to attack.

But which lab! Dr. Zenith changed his mind *again* and decided to teach in his own lab that day after all! I had to go back and tell everybody that plans had changed once more. The food really was going to be in Flash's lab. "Make up your mind!" their eyes said to me.

After Instrumental Chemistry, I ran down to finish laying out all the food. My hands shook as I opened packages of cups, and when I pulled apart the plastic, they flew into the air and across the floor. The adrenaline still surged through my system, and I took deep breaths to calm myself. Dr. Stillwell had given Mrs. Zenith an invitation call, and she and inorganic chemist Dr. Lisa helped me set out food, meats, cheese, sprouts, sliced cucumbers and other vegetables. The wonderful and forever-blessed Rebecca Furby arrived with chocolate covered Rice Crispy treats.

Flash left us to fill his vital role of delaying Dr. Zenith in the other physics classroom. Professors and students sprinkled in, and we encouraged them to start eating. I waited at the door and peered down the hall.

(Darn, Dr. Manchester was heading the wrong way! I hadn't gotten him the message in time. I dashed out and redirected him.)

As I stood at the door, I realized that I hadn't planned anything after that. What would we do once Dr. Zenith arrived at his party?

Stand around and eat? Would it be boring? I imagined possible ways to liven the luncheon when a voice wailed from down the hall, "They're coming!"

We all backed away - and just in time. Five black-clothed students exploded into the room, leaping and tearing to the back before the oncoming mass of Dr. Zenith. The astrophysicist landed on his knees and slid across the floor, flinging nerf darts at the retreating warriors in black.

Best entrance ever!

The ninjas laughed loudly, still half-hiding behind the lab tables, and everybody in the room laughed with them. I mean, we couldn't see the ninjas, because you can't *see* ninjas. But, we could hear them. With our imaginations.

There were no shouts of "Surprise!" None of that. Dr. Zenith didn't look surprised. He *did* look energized, and I shouldn't have feared the threat of dead air during any part of this glorious day. No, because this was a party for Dr. Zenith. Dr. Zenith the entertainer! And the Zenith Nebula took front stage during his special hour.

"I was standing in my lab talking to Dr. Gurden when I was attacked with a full onslaught," he told the room of people. "I fell to the ground to protect myself, and then one of them comes up and shoots me point blank! Who was that?"

Matthew had taken off his mask and was laughing loudly.

"Was that you!" Dr. Zenith grinned. "That was awesome!"

Kate and three others removed their masks as well, and the whole crowd happily munched away as Dr. Zenith described his day. He began with the letter on his door that started everything.

"But, I didn't read carefully enough, because the note contained a warning of danger, a literary device called 'foreshadowing,' which I enjoy very much. It said there were enemies I had to watch for. I didn't realize there would be actual enemies, so when they arrived, I didn't have my Thwacking Stick ready. I won't make that mistake again."

(Dr. Zenith, I love that you took that Thwacking Stick so seriously. Your melodrama at its best.)

One by one, the astrophysicist went through his birthday clues, describing the pains he'd taken to decipher each one.

"And then there was one that sent me to Dr. Gurden to find something about a gun to test the speed of sound. That sounded dangerous, but I went anyway. There were times during this morning when I thought, 'If I were watching me doing this, I'd be laughing so hard.'"

We all giggled along with him. It was pretty hilarious.

"You know," he said, "If a clue had told me, 'Crawl on your knees and quack like a duck,' I'd have done it. You do what the clue tells you to do!"

My mind widened. I'd been so afraid Dr. Zenith would consider the whole thing a waste of his time, I had not appreciated our power. What a marvelous attitude you have, Physics Professor!

"And under Bones' hat, I found a CD, and when I played the CD, it gave me songs - very good songs - that offered answers to another set of clues."

The song *Joy to the World*, by Three Dog Night provided him with the number "3." The song *Turn! Turn! Turn!* by the Byrds was taken from a certain book of the KJV Bible. *I Can't Help Myself (Sugar Pie, Honey Bunch)* required Dr. Zenith to know the artist, the Four Tops, which gave him the number "4," and Rod Stewart's version of *Rhythm of My Heart* came out a certain year.

"And when I filled in those clues, they provided me with numbers that applied to an equation. I did a little bit of math and filled in the blanks, and the answer gave me a license plate number. *My* license plate number. So I went out to my car, where I found a note on the windshield. It said, 'Good job, but this is not your license plate.'"

Laugher sparkled across the full room.

The clue at his license plate said, "THE TREES" and nothing more. These were the last two words in the phrase that led to the Lorax, but Dr. Zenith and his wife didn't understand that at first. They thought the clue referred to actual trees, so they hunted around the trees in the parking lot.

"There was a ribbon tied around one of the trees there, so we

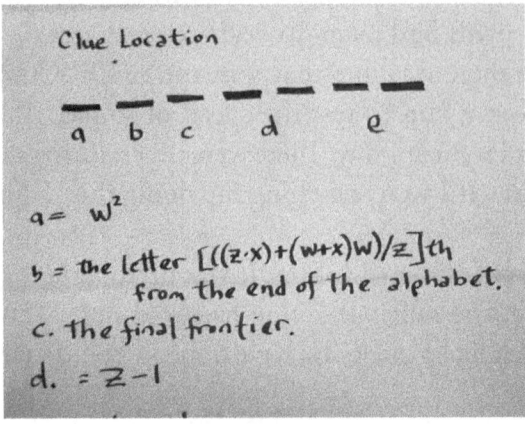

Figure 12: The bottom portion of the clue that decoded songs on the CD. (Haha. No, I will not give away Dr. Zenith's license plate info!)

tried to climb it. Nothing. There was nothing. Finally, in desperation I hunted down Dr. Stillwell for wisdom. Dr. Stillwell reminded me of the phrase in *The Lord of the Rings*, 'Speak Friend and Enter.' In Tolkien's story, they were looking for something complex, but they merely had to speak the word 'friend.' The answer was far simpler than I realized."

Dr. Zenith finally reached the part where he'd left Flash's lab to look up the phrase, "I speak for the trees." He himself had given Dr. Stillwell that Lorax, so he interrupted the geology professor at his lab again and said, "I speak for the trees, can I borrow your keys?" He then retrieved the weapon that was to have protected him from the enemies.

When Matthew presented Dr. Zenith with the ninja piñata, the professor refused to beat it to death. He wanted to keep it forever, so he stuck a knife into it from below and gutted it. Candy tumbled freely to the ground. That piñata still sits high on a shelf in his lab as far as I know.

Dr. Zenith loved his party despite the water, despite disaster. Dr. Manchester approached me while Kate cut the cake, and he gave me a long, warm embrace. "You did a good job."

Awww, Dr. Bob. You give such good hugs.

Dr. Zenith's wife and I stood together and watched while the festivities progressed. She whispered to me, "Thank you so much for doing this. He was so depressed when he came in this morning. He was expecting to have to clean up his lab."

"On his 40th birthday," I said.

She nodded earnestly. "This has been wonderful for him. Really wonderful. It absolutely made his day. It's the best birthday he's ever had."

Mrs. Zenith didn't know who I was. She'd simply helped me set up cups and food, so I don't think she ever blew my cover. I don't know if Dr. Zenith ever found out the part I played. He thanked Matthew and Kate, because they'd been dressed as ninjas, but he never thanked me. I'd been caught in the line of his other wall-eye when he'd thanked Flash, though, and that was enough for me. He didn't say a word directly to me, and I didn't say a word to him.

I'm glad. I liked it that way. He had a blast, and we had a blast, and I regard the day as a great triumph.

"I didn't understand the value of the Thwacking Stick," Dr. Zenith told the crowd. "I didn't understand, and so I was unarmed when the ninjas shot through the door. It's okay though, because the Thwacking Stick probably wouldn't have saved me anyway. It's like the Death Star. Awesome. Powerful. But no match for Luke Skywalker and the Rebel starfighters."

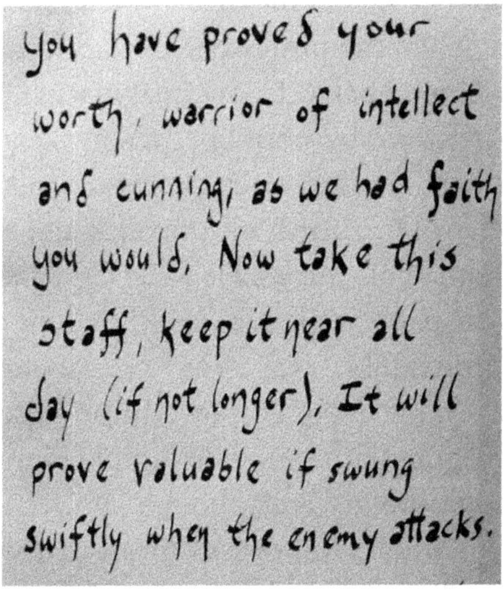
you have proved your
worth, warrior of intellect
and cunning, as we had faith
you would. Now take this
staff, keep it near all
day (if not longer). It will
prove valuable if swung
swiftly when the enemy attacks.

Figure 13: The note (left at Matthew's) originally meant to accompany our grand Thwacking Stick. Its replacement stated something equivalent.

Chapter 16
Dimples

Dr. Stillwell had dimples. Dr. Gurden had dimples. Dr. Manchester and the Lidge and Dr. Gromp and microbiology guru Dr. Park all sported charming dents in the sides of their faces. All of them. Even Crazy Ernie the great bryozoan expert, even he had dimples. I started noticing this particular pattern of cuteness among the science professors and wondered about it. At least half the science profs I knew were dimpled. My stepdaughter Amber had dimples. My own siblings Whitney and Lex both inherited our grandfather's single heavy crease in their good-looking right cheeks. I am jealous of them all.

At least I'm not alone. Dr. Zenith has no dimples either.

I have dreamed up a short movie in which Dr. Zenith is treated with indifference while the professors around him are adored and feted and ushered into a party. In my imaginary film, Dr. Zenith is blocked at the party door by stern linebacker-like guards who stare him down. Just then, another black professor walks up. The other black professor claps Dr. Zenith on the back and grins, displaying deep dimples in both of his cheeks. He walks right past the linebacker guards and into the party while Dr. Zenith is kept out. The cruelty of dimple discrimination!

I wish I had a dimple. I'd be all-powerful and take over the world if I'd inherited my grandfather's single dimple of magnificence. Dr. Zenith and I should be grateful, though. We're not grateful, but we should be. Our whole, full cheeks mean that we are physically superior to those dimpled mutants whose offices populated the science buildings.

Oh yes. Mutants. Dimples are caused by a division of the *zygomaticus major* facial muscle into two separate muscle bundles, generally the result of an error in genetic instructions during embryonic development. The very fact that there are so many dented cheeks in the world says something about the sexual selection advantages of this genetic defect. Who says that there are no beneficial mutations? Dimples are hot.

I was landed with a different sort of defect. I have freckles. Nordic hinterland pale skin spots. I produce melanin in sad little cluster-spit spatterings.

Again Dr. Zenith is the superior. He produces an abundance of melanin to protect his skin from ultraviolet cancer death.

"I love your freckles! They're so cute!" some girl told me at summer camp when I was 14. I'd never heard such loving foolishness. It's like I was tied to a wall in infancy and flicked repeatedly by five-year-olds with small paint brushes. I've been stained for life. The Cherokee blood in Randy didn't save our children's skin, but at least their specks are sequestered across the bridge of their noses where freckles belong. My children do have super cute freckles.

Genetic mutations have their place. While Dr. Zenith's dark skin protects him from the ultraviolet rays of the sun, my pale skin allows me to efficiently manufacture vitamin D in the slim light of northern winters. A spectrum of human beauty from Randy's dark hair and tanned skin to my pale hair and speckled face to Dr. Gurden's attractive dimpled cheeks results from genetic variations, which include mutations.

I'm melanin deficient across the board. I also have blonde hair and blue eyes. Blue eyes are caused by genetic tweaks that prevent melanin from invading the irises. There's no blue pigment coloring blue eyes. Rayleigh scattering of light makes eyes *look* blue, just like it makes the sky look blue. Green and hazel colorations are the result of some brown pigmentation in the iris, while clear blue eyes indicate little or no pigment at all.

Blue eyes sure are pretty, though. Blue eyes with black hair and dimples! Holy smokes. Shoot me now. Mutations can absolutely offer beneficial traits, whether by efficiently providing for increased vitamin D production during dark winters or by offering an advantage in winning a mate and making babies.

The mutation that causes sickle cell anemia has spread throughout portions of the African continent. Normal red blood cells are nice and round, but those with the sickle cell mutation are shaped like crescent moons, caused by a one-letter DNA substitution. One letter difference means a change in one amino acid, and that single mutation has caused multitudes of people to die in pain from misshapen red blood cells. These sickle-shaped cells stick together in the blood stream and clump up, and they fail to efficiently carry oxygen through the body. Sufferers are constantly oxygen deprived and usually die young.

However, the sickle cell mutation does have a benefit in regions with a high infection rate of malaria. Sickle-shaped red blood cells are resistant to malaria, offering an advantage to survival in areas where the virus kills multitudes each year. If a person receives a healthy hemoglobin gene from one parent and a sickle cell gene from the other, that person can still have normal red blood cells floating through the bloodstream alongside the sickle-shaped blood cells. A little mutation can offer the selective advantage of both getting oxygen to the cells and resisting malaria.

Remember, this mutation is caused because one letter is different in the hemoglobin gene. A thymine nucleotide base is substituted for an adenine nucleotide base in the beta subunit of the hemoglobin gene found on chromosome 11. That is, instead of an A, there's a T - and it codes for the amino acid valine instead of glutamic acid.

That's it! That's all. One letter.

The sickle cells greatly increase malarial resistance and the normal red blood cells efficiently carry oxygen. This survival advantage has allowed an otherwise destructive mutation to spread through the populations of the mid-African continent.

There are useful mutations. But, that's not really the issue.

Long-beaked finches on one island and fat-beaked finches on

another island, that's not the issue either. The issue isn't whether there are three-toed horses or umpteen species of elephants. The real issue is ... superheroes.

In 2004, the Associated Press reported that a little German boy had a genetic mutation that gave him muscles twice the size of other children.[1] The papers called the kid "superboy" and treated the child's genetic defect as a source of hope for people suffering from muscular dystrophy, a genetic condition that destroys the muscles of its victims. The child in Germany had the opposite condition, a rare genetic defect that caused the child's body to produce excessively dense muscles.

It sounds great. "Genetic mutation turns tot into superboy." Yeah! But, he won't be a superhero any minute. His genetic aberration didn't create anything new; instead, it caused a malfunction in the boy's ability to produce myostatin, a protein that limits the muscle growth in the body. Without proper myostatin production, his body will produce too much muscle. It might make him stronger than a normal boy, and he might look ripped without having to work at it, but he'll certainly be expensive to feed and he'll have to focus on stretching to keep those heavier muscles limber so they don't tear.

Actually, "too much muscle" really depends on the world in which you live. If there are low food supplies, then it's a bad thing to make big muscles and burn through calories. The people who survive in starvation situations are those whose bodies can store carbohydrates easily. Only when food is plentiful is it good to produce less myostatin and free muscles to grow more rapidly.

I haven't found any studies that show that superboy's mutation causes direct harm. It only appears to affect skeletal muscle and not heart muscle, and it may actually slow the effects of aging.[2] There are plenty of people who wish they had the same mutation: ("Oh noooo! I don't produce myostatin. Oh nooo! I look awesome without working at it! I need a support group to help me find a better place to tan!")

---

1 Genetic Mutation Turns Tot into Superboy. (2004, June 24). *Associated Press*.
2 e.g. Morissette, M. et al. (2009). Effects of Myostatin Deletion in Aging Mice. *Aging Cell*, 8(5): 573-583.

Still, the superboy's mutation didn't add something *new* that increased muscle production. No, a mechanism had broken, and his body no longer regulated muscle growth. Not only might that prove uncomfortable, he could be in trouble should a time of starvation hit.

From what I've seen, mutations don't add cool new code to DNA; they just cause existing function to break down. Superboy's genetic defect is similar to those whose pituitary glands don't properly limit growth hormone. These individuals keep growing larger and larger, until their bodies can't handle the weight. Sandy Allen (1955-2008), the tallest women in the world for 16 years at 7'7", would have kept growing had she not had surgery to remove a tumor on her pituitary gland. As it was, she had to use a wheel chair the last years of her life because her legs and back weren't holding up. The tallest man of our times, Robert Pershing Wadlow (1918 - 1940) died at the age of 22 when he was 8 ft 11.1 in tall. Few modern "giants" live long; most die young from medical conditions related to their height and weight.

At the time of this writing, the Human Genome Research Institute at the National Institutes of Health is busy filling a database with diseases caused by damaging mutations in the genetic code. There are plenty of neutral mutations, but a long list of dangerous or painful or inconvenient disorders are caused because somebody's DNA has an error.

Mutations don't create anything new: they only break down functions that once existed. This breakdown can be helpful in certain environments, but it doesn't change the fact that the mutation helped because something stopped working - not because something new was formed.

As far as I know, there are no databases collecting all the wonderful new superpowers that people have received because they are mutants. No databases describing the wings, the new organs, the fire breathing capability, the additional eyes or radar hearing of mutants. Those things happen only in the comic books.

Professor Charles Xavier and Magneto, Storm and Rogue are not real. They are inventions of the human imagination. A radioactive spider bite can turn Peter Parker into Spiderman, but in real life, that bite would have just left a big sore on his hand, and maybe he'd have gotten sick, and maybe his hand cells would have rotted away, and maybe he'd have died.

It's important for the body to limit growth hormone and muscle growth. Mutations that destroy the body's on-off switches don't make for healthy ... or adaptive...superheroes.

Chapter 17

The Hollow Man

Figure 14: Zeke, rocking his spring break, 2011.

Spring Break started a few hours after Dr. Zenith's birthday party. I took my Organic Chemistry exam and went home to pack. The children and I boarded a plane the next day to join my dad and stepmother down south at the beach for a week, their treat. The kids and I enjoyed the water and sun during the day, and Zeke drank many a virgin daiquiri in the kid's pool.

In the quiet, though, I worried about Dr. Stillwell. Maybe he'd ignored God's offers too many times. Maybe he'd turned his back so often that God had rejected him. I still felt joy in my spirit whenever I thanked God for loving him - God loved Dr. Stillwell - but I still worried about him.

Dr. Stillwell knew I believed that he mattered to God. This fact gave Dr. Stillwell comfort in hanging out with me, because I wasn't trying to fix him and convert him. When I'd wanted to criticize the good doctor, God told me to treat him with respect. When I'd wanted to argue with him, the Lord told me to keep my mouth shut. Apparently, my job was to enjoy the professor's company and plan trips to the Grand Canyon and be myself.

Scientists make predictions as part of the scientific method. We make observations and gather data. We form hypotheses. We then make predictions based on our hypotheses and develop experiments to test those predictions. If our predictions prove correct, we think we might be onto something and do more testing. I tend to apply the same sort of methodology when dealing with spiritual things. I'm not trying to publish papers, but I'm trying to figure things out, and the same principles work regardless.

I had been making observations. I observed that Dr. Stillwell felt empty. Sometimes, it overwhelmed me. I didn't want to tell him that he was a hollow shell, but I couldn't imagine I was the only one who felt it. There was something about the empty feeling he gave me, and I wanted to figure out exactly what caused it.

In my dream about Dr. Stillwell, I'd felt a deep comfort and warmth and open friendship. I'd felt fulfilled by it. Yet, in real life Dr. Stillwell didn't make me feel deeply comfortable and fulfilled inside. He soothed my sore heart, but he didn't give me a sense of contentment. I felt his emptiness. That dream wasn't about me, though. It was about him. And it made me think that perhaps... perhaps he got that deeply contented, comfortable, fulfilled feeling from *me*. I felt pretty sure that was it.

I think I have the Bible's backing on this. Jesus made a promise in John 7:38:

> *"He who believes in Me, as the Scripture has said, out of his heart will flow rivers of living water."*

Isaiah 55:1-2 says:

> *"Ho! Everyone who thirsts, Come to the waters; And you who have no money, Come, buy and eat. Yes, come, buy wine and milk Without money and without price. Why do you spend money for what is not bread, And your wages for what does not satisfy? Listen carefully to Me, and eat what is good, And let your soul delight itself in abundance.*

My hypothesis goes something like this: When I shake Dr. Stillwell's hand and it feels warm to me, then I know that my hand must feel cold to him. If Dr. Stillwell feels empty to me, then I must feel full to him. If I think he's dying inside, then perhaps he feels *life* inside of me.

I suggest that Isaiah 55 and John 7 describe something spiritually real. The Spirit of God fills us, feeding us with satisfying spiritual food and water. We have physical food and clothes in our rich western countries, but our spirits are starved. We're so used to it, we stop even feeling the hunger. I'm not spiritually hungry, though. I have a deep contentment in my soul, and even after Randy died, I had a firm sense that everything was going to be okay. No matter what craziness was going on in my life, no matter how angry or sad or upset I was emotionally, inside my innermost being I was solid. I do get sad and angry and aggravated, but deep inside I'm content. I believe the Spirit of God fills me, and my soul has delighted in abundance.

Perhaps Dr. Stillwell was spiritually empty, and he got a contented, warm and friendly feeling from me. He was starving to death, and I offered a lifeline.

That was my hypothesis, and I wasn't sure exactly how to test it. I decided I'd just have to talk to him about it.

There was another aspect to it. If Dr. Stillwell felt starved to me, that might mean something interesting. It might mean that Dr. Stillwell was alive inside. You can't be spiritually starved if you're dead. Maybe there was a sickly, gaunt life growing there, and God was going to water it and blossom it into something beautiful. That was my hope.

The kids and I played in the waves and splashed in the pool that week. We found a puffer fish washed onto the sand, and we climbed rosy granite rocks. We had a lovely time.

As the kids and I enjoyed our spring break, though, I worried about Dr. Stillwell. I hoped he hadn't rejected God one too many times.

March 19th, I wrote in my journal:

> When the kids went to sleep, I lay awake praying a little, and as I prayed for Dr. Stillwell, the burning came back. In my chest. I haven't felt the burning constantly like I did there for awhile, and I'm still trying to figure out what it means and what I'm s'posed to do when I have it. I kept praying for Dr. Stillwell…Sometimes I've wondered, because Dr. Stillwell's heart is so hard against God and the Christian God especially. But, the verse the Lord gave me last night as I prayed for him was John 17:24:
>
> *"Father, I desire that they also whom You gave Me may be with Me where I am…"*

I felt certain God had plans for Dr. Stillwell, this man He seemed to love so much. God wanted him with Him. But, it was obvious God Himself had to change Dr. Stillwell's heart, because it was something that I couldn't do.

Little did I recognize the greater importance of that trip to the Grand Canyon. The Lord would give me real insight about Dr. Stillwell's future on that trip, including a picture of the process that I believed would one day save his life.

Chapter 18

No Hammers!

```
To:
CC:
Date: Sun, 20 Mar 2011 19:33:19 -0400
Subject: I'm not in a trunk

Cher Professor,

I hope your vacation was productive and relaxing.

Just to let you know that as of right now, I have not been inconveniently placed
in the trunk of relatively inhospitable drug cartel lackeys. Nor have I been
swept out to sea. Nor have the condors made a light evening meal of me or the
children. Nope. I shall be around tomorrow (Tuesday) God bless it. Until
then, go ahead and enjoy your Monday unencumbered by visions of me in any of the
above predicaments.

I'll find out soon enough whether you bought the tickets.

AJ
```

When I returned to school on Tuesday, I discovered that Dr. Stillwell had not, in fact, had *not* purchased the airline tickets to Las Vegas for our Grand Canyon trip. Not for lack of effort. He'd spent two monstrously aggravated days working to reserve roundtrip tickets for the 10 of us going on the adventure, and the online ticket processing repeatedly dumped his efforts into the world wide trash. He went through the long process of entering our names and birth dates and all the piddly little required steps, and each time he pressed the "Purchase" button, he got an error message. There weren't enough seats available on any one plane. I'm sure he was ready to throw a rock hammer through his monitor.

Wednesday morning, he entered the GIS lab where a mess of us sat at computer, and he gave us the bad news.

"We're not going to be able to go to the Grand Canyon. The plane tickets have gone up to $500 each, and it's just too expensive. I think we'll do a trip to the northeast instead." No apology. No regret.

I stared at him in astonishment. After all our hours invested in the national parks of the Southwest, he wanted to flip and change our plans to Maine? In May? In the cold and rain?

"Nobody wants to go to Maine. Everybody wants to go to the Grand Canyon." I tried to calm my disappointment, but he had to know better!

Another student agreed. "We're all excited about the Grand Canyon. You have to work it out, Dr. Stillwell."

"And plane tickets should be only $350," I said. "I'll check again, but did you look at flights out of Baltimore?"

Everything would have been easier if we'd bought the tickets a month earlier as planned. But. The best laid plans of mice and men had failed as usual. The tickets had gotten expensive, and there weren't enough spots for us on the flights.

It didn't matter the odds, though. We had to beat the system and get everybody to and from Arizona. That afternoon, I spoke calm words of encouragement to our professor. To his great credit, Dr. Stillwell eased back in front of his computer, and I sat in my seat, and we found round trip tickets from Baltimore into Las Vegas for $375 each.

"Okay," I said. "Let's go through this one flight at a time. If we need to split up and take two different planes, I'll go with the students on the other plane."

"You will?" Dr. Stillwell asked with a flash of hope.

"Yeah. I'll be fine."

"It's not you I'm worried about," he grunted.

It took us two hours to buy those tickets. Two hours. Matthew's sister Megan watched my children that afternoon, and I'd promised to be there at 3:00 p.m. to pick them up so she could get work done. When 3:00 arrived, though, the good doctor and I remained stuck at the computer.

I watched from across Dr. Stillwell's desk as he typed, tense and silent. His whole body sat rigid as he tip-tapped through each step for the nth time.

"You want me to handle it?" I asked him.

He didn't answer.

"I can do it," I said. "You don't have to. Let me do it."

He didn't answer.

I glanced at his clock again. I needed to get the kids! The doctor had wrestled through the hardest parts, and he was now finalizing the last steps of the process. "You can finish up now, can't you?" I asked. "I have to go."

"No." Dr. Stillwell said it in a tone that meant, "Don't you dare." He didn't even move his eyes from the screen. I'd gotten him into this.

So I nestled my head into my arms on his desk and watched until he'd pushed every last necessary button. Twenty minutes later I raced to the Caerphilly farm to collect my children from Megan, where I didn't want to admit, "I'm late because I've been buying plane tickets." But it was done. The tickets were purchased. There was no changing plans now, and we were going on an adventure out west!

The next day, I sat in a favorite spot by the heaters at the end of the hall. I had settled on the floor and leaned against the big plastic recycling bin, comfortable and warm. I could rest in that toasty spot under the window and do homework away from the noise and bustle.

Dr. Stillwell approached me on his way to the parking lot. He apologized, "I'm sorry for keeping you last night. Did you get in trouble with Mrs. Caerphilly?"

"She was okay."

"Still, I'm sorry. I didn't want to get you into trouble."

"I could have left," I smiled. "I don't feel guilty for pushing you into buying the tickets, so we're even."

In answer, he ordered, "Walk with me." He often said that to me, and I hopped up and strolled down the hall with him toward the parking lot that held his Jeep.

"Did you have a good time at the beach?" he asked.

"Yes, yes I did."

"Did you drink a lot?"

"Oh, I had virgin daiquiris and piña coladas, so the kids could share them with me."

"You took the children?" he said, surprised.

"Well, of course. Who else am I going to go with?" I wondered if Dr. Stillwell thought I should have taken a human male to the beach. "Do you know how hard it is for me to find somebody who fits me?" I said to him. "Do you know how rare that is?"

I think he did.

"I dreamt about Randy again last night, and it was such a relief. In my dreams, I always forget that he's dead. He's always just been gone somewhere. I'm like, 'Would you please stop dying and coming back to life and just stay alive already?'"

Dr. Stillwell nodded thoughtfully. "Sometimes I wonder if any of this is real and if it isn't all in my head."

This was an old puzzlement among philosophy students. We all asked questions like that, and we got to the point where we realized we couldn't control other people with our minds. If we couldn't control them, they were probably autonomous beings themselves. That seemed likely.

"I'm not your imagination," I assured him. "I have a whole history that you didn't experience, that you don't know about."

"I do believe that you're one of the real ones."

"I am." I smiled at that. "I have thoughts you don't have."

"Are you sure?"

"Yes." I was certain of it.

Dr. Stillwell felt less empty than he had the last time I'd seen him. His heart felt softer. That was nice.

CHAPTER 19

QUANTUM ENTANGLEMENT

The more I work with the powers of Nature, the more I feel God's benevolence to man; the closer I am to the great truth that everything is dependent on the Eternal Creator and Sustainer...[1]

-Guglielmo Marconi, Nobel Laureate and inventor of the radio.

Let's go back to quantum mechanics for a bit. We know that electrons can act like both particles and waves, and that suggests (to me, at least) additional dimensions in the universe.

HEISENBERG'S UNCERTAINTY

The wave-particle duality of all objects in the universe leads to the uncertainty principle, introduced by Werner Heisenberg in 1927. Heisenberg. Heisenberg. Why do we know that name? Oh yes. *Breaking Bad* chemist Walter White chooses to go by that name, and for an understandable reason. Theoretical physicist Werner Heisenberg was brilliant.

Good old Werner asserted that it's impossible to know *both* the location and momentum of any single quantum particle (like electrons). Choose which one you want to know, location or

---

1   Marconi, M. C. (1995). Mio Marito Guglielmo. (R. Castoldi, Trans., p.244). Milano: Rizzoli.

momentum, because if you know one, you can't know the other. Heisenberg noted that we can measure a particle's location precisely, but when we do, we can't get a good read on its momentum. If we can measure its momentum, we've given up any hope of knowing exactly where it is. The more *certain* we are about its location, the less *certain* we are about its momentum and vice versa. That's the uncertainty principle.

Our measuring skills aren't the issue; it's a problem based in the very nature of reality. Waves don't have a specific location, because they're stretched out over space. But waves do have momentum. A short wavelength means a high momentum. A long wavelength means a l o w e r momentum. We can figure out how fast a wave is going, because momentum is related to both wavelength and velocity. But we can't nail down a wave to one location.

Figure 15: A particle has location. A wave has momentum.

This means that if we want to find the location of an electron, we can only look at it as a particle. We can't look at it as a wave. If we want to determine whether it goes through slit A or slit B, for instance, then we can only see it as a particle: the wave function has to break down. At the same time, if we're nailing down its location in a moment of time, then we can't determine how fast it's going.

The same basic rule applies to anything we can measure about quantum particles. For instance, we also can't know an electron's vertical spin if we know its horizontal spin. We only get one shot when we take pictures of quantum particles, and we can only capture one angle. And then there's the biggest issue: we change how particles behave just by looking at them. Quantum mechanics is nuts. It's not user-friendly like classical mechanics, not at all.

NEWTON'S MECHANICS

In classical Newtonian mechanics, there's no problem determining both the location and momentum of an object. Let's say we put a kid on a dance floor with two marbles. The child rolls one marble across the smooth floor at its twin marble. We know the mass of each marble, and we can measure their velocity. Since momentum is equal to mass x velocity, we can know their momentum at any moment. No problem. We can measure the first marble's velocity and the angle of impact, and we can calculate exactly where the two marbles will be at any particular second after smacking into each other. We have equations that allow us to figure it out. In fact, we can measure the velocity of one marble and use it to calculate the position of its twin marble at that moment. Easy breezy.

Law enforcement investigators do the same thing all the time. They can use the position of cars after an accident, along with things like the weight of the cars and skid marks, to figure out which drivers were speeding before the collision. The investigators can testify in court that Mr. Smith was driving 57 mph when he smashed into Mr. Johnson and not 35 mph like Mr. Smith claimed. Marbles and cars are easy subjects to handle in the big scheme of things.

We'd think it would be the same with "entangled" quantum particles - paired particles from the same source. If we can measure the position of an electron, we'd assume we could measure the momentum of its twin positron. We could then back-calculate and come up with the momentum of the paired particles at any given position. No, sir. It turns out that if researchers try to measure the *position* of one entangled particle, they can't measure the *momentum* of its partner. The wave function breaks down. It's as though the particles are in constant communication, no matter how far they are separated from each other, and they don't like being watched.

Electrons and photons and all the quatum particles know that the Heisenberg Uncertainty Principle is law, and they obey it.

Quantum Entanglement

It gets worse. Communication between paired quantum particles happens faster than light speed. Let me repeat that. Quantum particles communicate with each other – across space – faster than the speed of light. They know instantaneously what is happening to their twin after they've been separated from each other. It's real. The connection has been tested over and over.

The theory of special relativity tells us that nothing can travel faster than the speed of light, and Einstein thought the implications of quantum mechanics were horrible. He didn't like the idea of "spooky action at a distance" as he called it. In the famous EPR paper he wrote with Boris Podolsky and Nathan Rosen (Einstein, Podolsky, Rosen – EPR), Einstein insisted that the study of quantum mechanics was simply incomplete.[2] We didn't understand what was really happening, and it only *looked* like paired particles were communicating. EPR suggested that quantum particles had what they called "hidden variables" – a sort of particle DNA. These hidden variables told the entangled particles what to do in advance, and it only made them appear like they were talking on little quantum walkie talkies.

I feel Einstein's pain. I don't want it to matter whether we are *looking* at a particle or not. I want it to act like a particle regardless, just because that's what it *is*.

But we don't always get what we want. There are no hidden variables. Electrons and photons and other quanta really do read each other's minds.

In 1964, John Stewart Bell devised an experiment to test quantum communication.[3] I won't open that can here; it's brilliant, but tedious. The important thing is the conclusion: Bell showed, based on the statistics of their behavior, that there aren't any hidden variables and the particles *are actually communicating* with each other. Whether Einstein liked the idea, spooky action at a distance

---

2   Einstein, A; Podolsky, B; and Rosen, N. (1935). Can Quantum-Mechanical Description of Physical Reality be Considered Complete? *Physical Review*, 47 (10): 777–780.
3   Bell, J. (1964). On the Einstein-Podolsky-Rosen Paradox. *Physics*, 1: 195–200.

was the winner.

Starting with John Clauser in 1978[4] and Alain Aspect in 1982,[5] a variety of researchers have since demonstrated the reliability of Bell's statistics. Each particle knows what is being done to its partner, and they know instantaneously. Hello quantum computing.

In fact, all particles in the universe might be connected to each other, back there behind the scenes, where all the tassels are woven into the rug. Wow wow wow.

DIMENSIONS AGAIN

Einstein long ago told us that there are more than three dimensions. Einstein's theory of general relativity tells us that *time* is the fourth dimension. Space-time is a fabric that can stretch and shrink and bend, and Einstein described gravity as a warping of that fabric. Even from general relativity, though, we can argue there's at least five dimensions. If space-time bends and warps, it has to have an additional dimension to bend *into*. Boom, five dimensions right there.

Mathematicians use formulas to describe reality, and those formulas can be beautiful or they can be awful and clumsy. It turns out that adding dimensions makes the universe more elegant mathematically.

In the 1920s, the Kaluza-Klein theory added a fifth dimension to the mix. Theodor Kaluza and Oskar Klein unified gravity and electromagnetism by connecting electromagnetism (think ultraviolet waves, radio waves, microwaves) to gravity curled up in the fifth dimension. The equations worked.

Do you see the pattern? Einstein added another dimension, and that explained gravity. Kaluza and Kline added another dimension, and that explained gravity and electromagnetism, all in pretty mathematical terms.

In the 1980s, Edward Witten's "M-Theory" made the case for

---

4 Clauser, J. F., & Shimony, A. (1978). Bell's theorem. Experimental tests and implications. *Reports on Progress in Physics*, 41(12), 1881.
5 Aspect, A.; Grangier, P; Roger, G, (1982). Experimental Realization of Einstein-Podolsky-Rosen-Bohm Gedankenexperiment: A New Violation of Bell's Inequalities. *Physical Review Letters*, 49(2): 91–94.

11 dimensions, six of which were wound up as one-dimensional vibrating superstrings. Today, theoretical physicists are searching for the Theory of Everything, mathematical equations that will unify all the forces of nature. Give them time.

Okay, that was the hardest part to picture. Maybe.

Zero-Point Energy

We also know that that there is incredible, unbelievable energy in every square centimeter of "empty" space. When there are no atoms, no nitrogen or oxygen molecules bouncing around, there is still a vast amount of energy in the fabric of space-time itself. This is called zero-point energy (ZPE).

Remember the bad guy from the Pixar movie *The Incredibles*? The red-headed evil genius Syndrome uses a beam to capture Mr. Incredible, saying, "It's cool, huh? Zero-point energy. I… I saved the best inventions for myself." Mr. Incredible's nemesis uses ZPE to freeze Bob Parr and his superhero family, holding them in mid-air so that only their eyeballs can move. It *is* cool, Syndrome. Evil geniuses sometimes develop sweet ideas.

ZPE is credited with making helium atoms buzz even at a temperature of absolute zero. Atoms of ice frozen in a crystal lattice still oscillate. They refuse to sit still. See, if they stopped moving altogether, then we could measure their momentum as zero, and we'd measure their position, and we'd beat Heisenberg's uncertainty principle. HaHA! However, subatomic particles all know the law, which requires them to eternally bounce around like two-year-old children.

In real life, It's unlikely that ZPE could ever be used to freeze Mr. Incredible. Each square centimeter of space holds the energy of ZPE, and while it's unbelievably powerful, we can't tap into it. The very reason it doesn't fry us all is the fact that its potency is *balanced* all around us. Like air pressure. We're not squished by air pressure, because it's pushing in all directions, cancelling itself out. If you suck all the air out of a plastic bottle, it squishes. Why? Because the air inside and the air outside are no longer balanced. Balance is important.

The ZPE is likely what holds electrons balanced in their atomic orbits, keeping them from shooting off into space or from spiraling into the atom's nucleus. Thanks for that insight, Harold Puthoff.[6] Empty space isn't empty at all. It is veritably pulsing with life and power, and subatomic particles regularly pop in and out of existence like bubbles of foam. So fantastic.

What does that mean for reality? How are entangled particles passing notes across space faster than the speed of light? How are they popping in and out of existence? How is that possible? I think it's because *we can't see what's going on* in those additional seven dimensions the physicists tell us are there. The particles are talking to each other behind the curtain. They're popping back and forth through the curtain. Now you see me. Now you don't.

Einstein made a cool - or disconcerting - conclusion about what all this means for the world we see with our physical eyes. He said, "Reality is merely an illusion, albeit a very persistent one."

And then Woody Allen joked, "If everything is an illusion and nothing exists, I definitely overpaid for my carpet."

The real problem isn't that nothing exists, though. The problem is that *more* things exist. The world goes deeper than the carpet tassels, even while we live out here on the surface.

Here's what I know. I know that I can feel things about people when I'm not with them. I've sensed things about Dr. Stillwell from clear across the country. Whatever causes entangled particles to connect, there's something going on behind the scenes that tells me that I'm also connected to something bigger.

What's more, the direction I receive is personal. It tells me, "Yes," and it tells me, "No." It tells me, "Wait," and it says, "Okay, now go."

---

6   Puthoff, H. (1987). Ground state of hydrogen as a zero-point-fluctuation-determined state. *Physical Review D Phys. Rev. D*, 35(10), 3266-3269.

Chapter 20

Keep Smiling

AJ: We'll get Dr. Stillwell a tranquilizer gun for his birthday!
Hannah: If we can figure out when it is.
AJ: Oh, we'll figure it out.
Hannah: June 6. D-Day.
AJ: D-Day.
Dr. S: (A slow, knowing nod.)
AJ: We're going to get you a tranquilizer gun for your birthday, Dr. Stillwell.
Jared: I think, that if we got him a tranquilizer gun, it's *you* he'd be most likely to use it on, Amy Joy.

Dr. Stillwell was a straight-edged old guy, and I appreciated him for it. It was such a relief to be able to hang out with him and relax. Sometimes he treated me like his adult daughter, and sometimes he treated me like his 12-year-old daughter. And sometimes he treated me like his partner in crime.

"There's a lot to you that's still 12-years-old," he said one day.

"I know," I said. "I think it's the 12-year-old part of you that gets along with me so well."

He didn't disagree. "If I'd grown up all the way, I'd still be in the oil industry making lots of money."

I didn't understand what he saw in me, though. He really set me aside and treated me differently than the other students. It bothered me. I wanted to explain things I never said in real life, things about my childhood. I wanted to tell him about breaking my thumb and never asking my mother to take me to the doctor. I wanted to tell him that God loved him and didn't want me to harm him.

I did explain to him one day, "I just want somebody to yell at me to get down from hillsides."

"I don't do that, do I?" he said.

"What? You've done it every time!" He absolutely remembered yelling at me to climb down. He loved playing ignorant.

Late in the Spring 2011 semester, I slumped onto the floor out in the hallway feeling grieved. I had purposely arrived at school early so I could have a talk, a real talk with this professor who'd drawn me so close. He didn't know that I rarely bonded with anybody. I had moved so many times in my life, I'd left so many people, I'd lost so much, that I didn't even have the ability to bond with most people. And yet, he'd managed to get through a gap in my leather armor and into my sore heart, and that meant he could actually hurt me.

I wanted some answers from him. What was the deal? Why did he drag me into his office, why was I important to him? What were his intentions? I had my dream and my guesses on the matter, but I didn't have actual answers from his own mouth. I trusted in God's leading, but I also wanted to hear what Dr. S. had to say for himself.

I sat against the wall by Dr. Stillwell's office door, waiting. Dr. A. came along first, and of course he had priority over me. Just as Dr. A. left, Dr. Zenith invaded. When Dr. Zenith finally strode out, I jumped up to walk into the office, but the astrophysicist closed the door in my face. "He's on the phone with the dean."

I slid down again and leaned against the radiator under the window. Dr. Zenith hovered above me, not leaving right away.

"I need to talk to him," I grumbled.

He actually paid attention to me. He stood over me as I sat on the hallway floor with my back against the cold wall heater. "You look sad," he said.

I watched him back without responding. We were at peace, Dr. Zenith and I, but I wasn't ready to expose the raw places of my soul to him.

Dr. Zenith seemed genuine at that moment, almost kind. He even began to tease me lightly. I wasn't paying attention to him, but it wiggled into my mind that the pompous astrophysicist was trying to cheer me up. I looked up at him, aware that a new level had been reached in our relationship. We'd come a long way, he and I.

Dr. Stillwell opened the door and walked out. The good doctor tossed a piece of chocolate at me, and it would have been merry had my heart not ached.

Dr. Zenith continued on down the hall to his own office while I pulled myself off the floor and tucked the chocolate into my jacket pocket. I dropped into my seat by Dr. Stillwell's desk, and the geology professor returned to his side. He behaved as though I belonged there, as though that seat had been placed beside his door just for me.

He talked for a bit about school stuff and trip stuff, and while he talked, I sat and played with a penny. I passed that circle of copper metal back and forth between each hand, methodically turning it around and around in my fingers.

"Um," I started, trying to think of what to say, staring at my penny. I needed to get to the point. "I have something bothering me, and I don't know whether it's better to bring it up or to leave it alone."

Dr. S. said nothing. He just looked at me, waiting.

My eyes brimmed with tears. I had never bawled in his office, and I didn't want to start now.

"You… you touch a very sore spot in my heart."

His eyebrows rose. "I'm sorry."

"No… no…" I reassured him, my stupid eyes stinging hot and red. That was the problem with trying to say these things. Anytime I practiced in the mirror, my irises glowed green because of the annoying red, swollen emotion in my face.

I shoved the tears out of my eyes and kept at it. "It's okay. It's just… There's this big void that's been there all these years – for decades – and you fill it very well. And I'm *glad* you drag me into your office. I don't have to have a reason for you to drag me in. I'm glad you enjoy my company. I just… I just don't know what I do for you." I looked earnestly at him. "What do I offer you?"

"You and I have a trip to plan…" Dr. Stillwell started to say. Then he stopped. He knew that's not what I was getting at. "And you have my bizarre sense of humor, and you have a great smile. Do you think I get smiles in here every morning?"

I pushed away at my rebellious tear ducts and laughed. "Here I am, crying in your office. That's what you try to avoid, people crying in your office."

He agreed, "Yes I do."

"Thank you," I said finally. "I'm grateful that you want me around. And I'm grateful to you that you aren't stupid. So many men are stupid."

He understood what I meant, and as I said it, his eyes softened. "I'm going to be honest with you, more honest than I often am." He wasn't wearing his glasses, so I was able to see his eyes without any thick glass barrier. His quiet, intelligent eyes gazed at me openly. "I care about a lot of people, but I care about you. And why would I do something that would hurt you if I care about you?"

That's why I liked him so much. I nodded in appreciation. "A lot of people don't get that."

"And you're not stupid either," he pointed out.

"No, I'm not."

I heaved in air and tried to get myself together.

"So, what's going on with you these days, Amy Joy?" The good doctor smiled at me, trying to move on to a less charged subject. "I know you stay busy."

I sighed. "Well, they're filming my screenplay, and I'm going out Wednesday night to act in one of the scenes. So, that should be fun."

"Who's making it?" he asked.

"The single-camera production class. And my brother Baron is finally going to illustrate one of my books so I can try to get it pub-"

"See!" he interrupted me. "How many people do I have coming in here who say, 'Oh, they're making my movie' or 'they're illustrating my book!'" He said it with mock disgust that made me grin. *I* knew that unskilled students were practicing with the script I'd written for screenwriting class the previous spring. There was no real glory in any of it. But, it pleased old Dr. Stillwell.

"Just smile and enjoy life, Amy Joy," Dr. Stillwell said. "Keep smiling."

It was an important conversation. It was important to establish that neither of us were stupid. So many times in the past five years we've been like friends, or like a father and daughter, and yet, so many times during the past five years we've been like two kids who lost our childhoods and started building a fort in the yard decades out of time.

He didn't really answer the question though. He didn't. I don't think he even knew what it was himself. Not really. What is it he gets from me? Thousands of students have walked through his classroom. I recognize that I'm brilliant and charming (and stubborn and aggravating and less than respectful). I'm a cute girl. I have freckles. All that stuff. But he enjoyed my company even when I sat quietly and said nothing at all. He liked having me in his office, reading my geology book. What unique thing did I offer him?

We took a class trip to cold, rainy Maine in May of 2012, and there I warned him, "I'm moving out to Idaho. I'm going to leave."

He rejected that idea. "No you're not."

Of course I moved. I had a job offer from a geochemistry lab in my Inland Northwest homeland. I bought a house. I went on with my life.

But, if I haven't called him for a long time, Dr. S. doesn't easily let me off the phone. It's the summer of 2015 as I write this, and I recently visited him in West Virginia. We spent a long time talking, and he reached across his desk and grabbed the top of my hand. "I will call you more often," he said earnestly.

"You don't have to," I told him. "I'm not a burden."

"No. I will. I promise I will."

Dr. Stillwell loves his wife, his cheerful delightful little wife. His eyes light up when he talks about her. He loves his stepchildren. He breaks my heart, though. What is this wound inside him, this hole that I fill for him? Where does the bitterness, the sadness come from? I'm still not sure about the full answer to those questions.

Within a month of our turning-penny talk, God gave me a prophecy about Dr. Stillwell. There was a reason He'd stuck us together in the first place after all, and it went beyond wounded hearts and friendship. I believe Dr. Stillwell was in danger of dying, and God planned to save his life.

Chapter 21

Scroggin and Ferraris

Dr. Stillwell didn't want to hear about my superpowers, and I didn't really want to tell him. But, it did slip out. I'd told him about the fire. As the months went on, I opened up to him about everything: the dream(s) I had that warned me in advance what was going on with him, or the fact that I knew when he wanted to see me and when he didn't have time.

I told him that stuff, because I can't keep my mouth shut like a wise person. Yes. I told my geology professor that I have spiritual superpowers. The fact that it's true doesn't mean I have to blab it out.

He got super angry at me when I finally told him that he felt empty to me. He thought I was trying to manipulate him, that I was suggesting he was inferior because he didn't share my faith.

"That's why I didn't want to tell you," I declared in his back room during the fall of 2011. "I didn't want to tell you that!"

He finally softened. "Well. What do you think it is?"

"The Holy Spirit," I shrugged. That's what I honestly thought, so I said it.

Most of the time, Dr. Stillwell rejected my conclusions and sought alternate explanations, but I'd been listening to the Holy Spirit's guidance for as long as I'd known how to do it. When people tell me, "I love your spirit," they might mean that I have a lot of energy and humor, but I've asked the Lord to splash other people through me, and I'm confident He does it.

Jesus says in Luke 11:10-13 that the Father will give the Holy Spirit to anybody who asks for it. He loves to give good gifts to His children, which means even me.

Dr. Stillwell doesn't like my explanation, so he rejects it. Instead, he accuses me of having weak logic when it comes to matters of faith. "You're very intelligent, Amy Joy, but I think you break down."

No. I don't. I work hard to be honest with myself.

That's why this issue about evolution is so important. Dr. S. thinks that we evolved as a matter of course, and God isn't necessary. I don't have a chip on my shoulder. I don't want to pick fights, but it's important to figure out who is actually breaking down.

I took biology at Gonzaga Prep in Spokane, Washington at the impressionable age of 15. Mr. Scroggin liked to stand behind me at the beginning of class and drop his massive bear hands onto my shoulders. He'd shake me, "How's my little Truman friend doing today?"

I'd cringe. Please don't do that, Mr. Scroggin. Don't single me out.

Mr. Scroggin would then lumber to the front of the room and attack the board by chalking out formulas and cycles. Sadly, he didn't warn us about the topic for the day, so we never knew whether he was talking about photosynthesis or the Kennedy assassination, and we all sat, wide-eyed and lost. As soon as possible, one of the boys asked, "Hey Mr. Scroggin, did you go hunting this weekend?" That almost always saved us, and Mr. Scroggin fell into telling stories about his dogs or knocking over outhouses as a kid.

Old Rod Scroggin did manage to get a few biological principles across. As I told Dr. Stillwell, we were forced to chant the phrase, "Ontogeny recapitulates phylogeny." Ernst Haeckel's belief that the embryos of all critters went through a fish stage and a chicken stage (et cetera) had long been disproven, but the idea lingered that the stages of embryonic development demonstrated each species' evolutionary history.

I managed to earn an A in Biology. I learned a lot in the end, because I decided to read the textbook. I poured over each chapter, reading and re-reading every paragraph until I understood the concepts. When we took a standardized biology test at the end of the year, I earned the second highest score in our Jesuit prep school.

Despite Mr. Scroggin, class wasn't entirely useless, because we dissected a lot of creatures. The aroma of fish in formaldehyde ranks right up there with swollen, maggot-ridden roadkill and outdated old lady perfume, but dissecting those animals gave me solid insights into anatomy. When we examined the insides of a cadaver at Eastern Washington University on a field trip, the heart and liver and kidneys matched the smaller-scale versions I'd seen inside the frog I'd opened up. Frog liver = small. Human liver = big.

The more I studied biological systems at the age of 15, the more impressed I became with the brilliance of it all. From single cells to flatworms to amphibians to mammals, life required a tremendous amount of interrelated processes, like cogworks, all cooperating and depending on one another. When systems were considered, astonishment grasped my lungs and sucked the air out of them. In order to survive, a creature didn't need just a beating heart, it required the veins in place to hold the pumped blood and lungs ready to absorb oxygen from the air and hemoglobin in red blood cells to carry that oxygen. It needed an immune system ready to battle off invaders and a digestive system to process nutrients and a pancreas for insulin to pull sugar out of the blood and nerves to spread their fingers into every tiny part of the body and send the messages to keep the body going. All were necessary, and all had to be functioning at the same time.

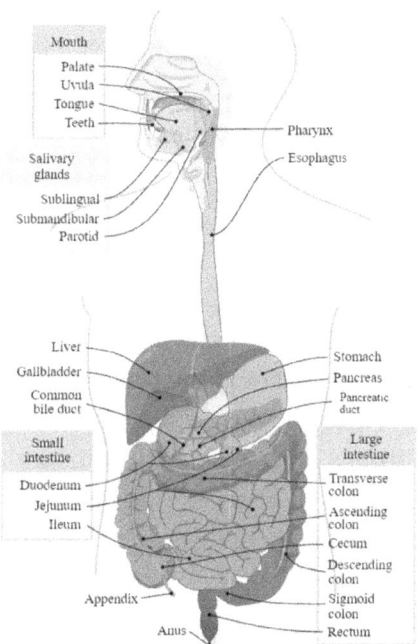

It didn't matter how much Mr. Scroggin told us it had evolved that way. The more I learned, the more skeptical I

Figure 16: A diagram of the human digestive system, care of "LadyOfHats" Mariana Ruiz Villarreal.

became that any of it developed through step-by-step processes. The interdependent parts of systems would have had to form all at once. What value did the heart have without arteries or hemoglobin or cells begging for oxygen? How did the bones function without cartilage to protect the joints and tendons and muscles to move them? Muscles needed both actin and myosin, the push and the pull. They needed calcium and potassium ions pumping signals to nerves.

All these marvels magically came into being through natural processes? They seriously expected me to believe that?

Mr. Scroggin's noise about ontogeny recapitulates phylogeny and peppered moths didn't impress me at all, but I assumed that people in the biological community were smart and had a clue, so I tried to figure out why people *did* think it all evolved. We were told all the time, "Evolution has been proven." Biologists seemed to agree, but I couldn't figure out what had them so convinced.

In his 2002 book *The Constants Of Nature*, John Barrow notes that the universe itself looks like it was designed with the purpose of sustaining life. Barrow is a bit of an expert on what's called "the anthropic principle," the idea that the universe looks like it was engineered for the benefit of higher lifeforms, like humans. He and Frank Tipler even wrote a book on it.[1] Barrow recognizes that the universe appears designed for our existence, but he says:

> [A]nd the plainness of the evidence for such design meant that there had to be a Designer.
>
> As it stands this ancient argument was difficult to refute by means of scientific facts. And it was always persuasive for those who were not scientists. After all, there are remarkable adaptations between living things and their environments all over the natural world…But scientists are never very impressed by such arguments unless they can provide a better explanation. And so it was with the Design Argument.… It was only overthrown as a serious explanation of the existence of

---

1 Barrow, J., & Tipler, F. (1986). *The Anthropic Cosmological Principle*, Oxford, Oxford University Press.

complexity in Nature when a better explanation came along… The better explanation was by means of evolution by natural selection, which showed how living things can become well adapted to their environments over the course of time under a very wide range of circumstances… Complexity could develop from simplicity without direct Divine intervention.[2]

At the age of 15, it was easy to look at the wonder of life and see Design. After all, "it was always persuasive for those who were not scientists." With those words, Barrow summarized the much-repeated view of the scientific establishment that if I were an educated scientist, fully informed, able to understand complex biochemistry and developmental biology, then I would appreciate how completely and thoroughly the grand theory of Evolution had been proven.

Even back at age 15, I knew I had to be missing something. In all honesty, really and truly, I couldn't see how all the creatures on the planet evolved from bacteria, or how bacteria had evolved in the first place.

Don't misunderstand me. I didn't have a problem with survival of the fittest and natural selection and the constant changing and adapting of species to new environments. Various cat and elephant and horse species had obviously diverged over the ages. Certain species had passed on some genes in one location, but not in another.

But there was this massive, looming hole in the whole thing. Barrow made that leap in his 2002 book. The scientific community since Darwin seemed to have made the same leap in logic. Yes, evolution by natural selection gives animals the ability to fill every kind of niche and adapt to their environments. But when Barrow wrote that natural selection meant "complexity could develop from simplicity," he asked me to hop over the Grand Canyon with him, and I couldn't do it.

All over creation, we see species straining out certain genes or expressing this gene more while expressing that gene less, but these

---

2   Barrow, J. (2002). *The Constants of Nature: From Alpha to Omega--the Numbers that Encode the Deepest Secrets of the Universe* (p. 158), New York: Pantheon Books. Italics are in the original.

things *do not create new function*. They don't. There had to have been the development of new function for any of us to evolve from amino acids, but I never saw anything *new* evolving in real life. I only saw genetic information being strained out.

My freckles and Dr. Stillwell's dimples and every beneficial mutation I've ever seen are the result of a genetic *loss*. Peppered moths and malarial resistance due to sickle cell anemia don't create anything new. They are the result of mutations that destroy function that once existed. Not one of them creates new machinery. The radioactive wave of love from the Chernobyl nuclear meltdown produced tumors and sickness and death, not glorious new life forms. Not superheroes.

It's even worse now that we know about epigenetics. Epigenetics is the study of how the body turns genes off and on. Identical twins are born with the same genetic code, but some genes might be used more in one twin than in another, and so we see differences that help us to tell them apart as they grow. Those epigenetic changes are built into the system already. We're engineered to adapt to changing environments, to automatically adjust our metabolism in colder weather or dial up our tanned skin in the summer. Epigenetic changes turn white fat cells into thermogenic "beige" cells, which release heat energy in the winter. These adaptations take place automatically, but only because we're pre-programmed for them.

The function is already there. Evolution can only act on machines that already exist.

One of my goals as a biochemistry major was to answer the question I'd puzzled over: why did biologists think that microbes-to-man evolution had been proven? I assumed there were better answers than the ones I'd already heard, but I couldn't see them for myself. I had to go hunting. I had to figure out why on earth all these obviously intelligent people believed Design wasn't necessary.

The subject came up in the spring of 2012 on a geology day-trip for new students. Dr. Stillwell drove the school van and I sat shotgun, and we had fun exchanging stories with the teacher education students in the back. At some point between Blackwater Falls and Seneca Rocks, somebody in the back laughed at the ignorance of creationists. We zoomed along in the van after the derisive remark, and nobody said anything for a minute. I didn't intend to say anything. I had no desire. I had finally figured out some of the major reasons biologists believed Evolution created us all.

First, all living things have similar blueprints. Frogs have a liver and kidneys and a pancreas to produce insulin and a brain to send electrical signals - just like we do. All life on the planet has DNA just like we do. All DNA uses nucleotides made of the same elements. The cells of all eukaryotes on Earth have a nucleus and ribosomes and mitochondria. All of them! We all use the same kinds of hormones and the same kinds of transcription factors. The hormones and transcription factors have slight differences from creature to creature, but, we're all an awful lot alike.

In 2010, biogeochemists suggested that the extremophile bacterium GFAJ-1 used arsenic instead of phosphorous in its DNA,[3] but it turns out this bacterium still uses phosphorous over arsenic anytime it can.[4] All the rest of life uses phosphorous.

There are entire sections of DNA code that match between widely separated groups of creatures. Consider the DNA of chloroplasts, which keep plants green. Plants have three different sets of DNA; they have DNA in their nucleus, in their mitochondria, and in their chloroplasts. There are a wide variety of differences in nuclear DNA and even mitochondrial DNA, but chloroplast DNA is remarkably similar between land plants, from mosses to oak trees. Chloroplast DNA is also highly similar between land plants and charophycean green algae.[5]

This observation gives the impression that all land plants are related to each other, that they all descend from a chloroplast-producing ancestor (much like charophycean green algae).[6] And

---

3   Wolfe-Simon, F. et al. (2010). A Bacterium That Can Grow by Using Arsenic Instead of Phosphorus. *Science*, 332 (6034): 1163–1166.
4   Elias, M., et al. (2012). The Molecular Basis of Phosphate Discrimination in Arsenate-Rich Environments. *Nature*, 491:134-137.
5   Turmel, M. (2005). The Complete Chloroplast DNA Sequences of the Charophycean Green Algae *Staurastrum* and *Zygnema* Reveal that the Chloroplast Genome Underwent Extensive Changes During the Evolution of the Zygnematales. *BMC Biology*, 3:22.
6   Turmel, M. et al. (2006). The Chloroplast Genome Sequence of *Chara vulgaris* Sheds New Light into the Closest Green Algal Relatives of Land Plants. *Molecular Biology and Evolution*, 23(6):1324-1338.

so with animals. It would make sense that fish and worms and humans all have hearts and muscles and hemoglobin if those wonders developed early in the history of life and simply modified a bit over the millions of years of evolution.

What's more, biological processes are automated! Chemical reactions automatically take place in our bodies every day without any magic involved. Our mitochondria offer us the electron transport chain to produce ATP from glucose over and over and over a vast multitude of times each day, giving us the energy we need to operate. Throughout our bodies, one reaction causes the next reaction, which causes the next, a circle of life that continues steadily to make sure that our cells grow and develop, our hearts pump, and hemoglobin hands off oxygen to our cells. It's all a whirring, well-oiled machine.

And recall, there really are different critters found in different geological layers.

Bottom line, I've found that:
1) All creatures on the planet share the same basic ingredients.
2) It's all automated.
3) Paleontologists find different creatures in different geological layers.

These well-established bodies of knowledge present a pretty good circumstantial case for the evolution of all life from common origins in the past.

So, I had reached a puzzlement.

I rode in the van next to Dr. Stillwell and pondered while he drove down that hilly West Virginian highway. The young woman in the back had expressed her scorn for creationists, and she was free to do so. I didn't feel like arguing.

Then, I thought, "Shoot." Maybe I should offer a bit of additional perspective. It wasn't right that the students should laugh at creationists, because there *is* another side to the issue.

"You know," I said after a long pause. "We all use the same proteins, from amoebas to humans. We're all made from the same stuff."

Agreement rustled around the van.

"But, we have a problem. Because DNA can only go so far. Peppered moths didn't create new DNA to produce the pigment that camouflaged them on dark, dirty trees. They didn't fly around screaming, 'Ahhh!! We have to DNA-up some darker pigment or we'll all die!!' Nobody thinks they did. The moths with the best coloring for the environment were the moths that survived, but the DNA for the pigment - or lack thereof - already existed in the gene pool."

I went on. "In order for true evolutionary change to take place, there have to be useful mutations, and we don't see the truly creative mutations we'd need." I reminded Dr. Stillwell, "Any mutation you see breaks down something that was once functioning, and it causes *less* function. Even certain mutations like sickle cell anemia - it might make Africans less susceptible to malaria, but nobody can say that sickle cell anemia is an improvement.

"It's like… like your Ferrari is racing toward a cliff. If the car's transmission falls out, it saves you from going over the cliff. But nobody is going to argue that a broken Ferrari is better than a running one."

If evolution made all of us from scratch, creative mutations might not be common, but we should at least see *some*.

"We don't have those kinds of beneficial mutations," I shrugged to the van of education majors. "The Ferrari that broke down at the cliff edge survived because it stopped working right. It didn't sprout wings and fly off the cliff; it just broke down. You can't find a mutation - and I've looked - that produces new function. Mutations aren't a good mechanism. They always break down or change up the function that's already there."

Dr. Stillwell didn't express offense at what I'd said. We all recognized the problem.

Then he said something very revealing. "But, we can't just give up. We can't just stop looking. We have to believe the answer is out there. We have to keep searching, expecting to find a mechanism one day. That's what science does. It keeps on hunting and searching until it finds the answers."

I stared out the windshield in front of me, thinking about what he'd said.

On one hand, he's right. We'd still be blaming lightning on Zeus if we didn't understand electrostatics. We'd be blaming diseases on bad spirits if we hadn't seen microscopic bugs blooming under magnification lenses. We need people to keep hunting behind every toadstool for more understanding about how biological processes work. We can't give up.

On the other hand, there's a heck of a lot of *faith* involved in what Dr. Stillwell said.

This is a deep issue. Books upon books have been written about the minute details involved in the questions of life's origins, and I've been engaged in a brutal, muddy wrestling match over the conundrums that rise before me like Macbeth's witches. It's cruel to overwhelm you with them, dear reader, in the middle of what is supposed to be a story, so I won't. I will relegate the more technical matters to the Appendix, but I hope you read those additional pages. It's important that we hunt down answers to these questions.

Dr. Stillwell had faith that we'd find a novel function produced by natural selection, that we'd be able to watch evolution in its ability to build and grow structures - contrary to everything we observe about the way mutations work. Not evolutionary storytelling, but actual observations. Dr. Stillwell had faith that evolution could have produced wings and brains. He trusted that marvels like speech or spider spinning skills or the woodpecker's shock absorption system all developed from small changes over time. He had faith that evolution was able to produce the circulatory system from scratch way back near the beginning of eukaryotic life.

There was good circumstantial evidence for the grand theory of Evolution, but it was still circumstantial. When I got down to the bolts and nuts, I *still* couldn't see how it could possibly work.

We all had the same ingredients, but so what? So do nuts and bolts. We use screws and nails and staples made from the same materials, yet they were all designed on purpose for specific reasons.

How did those ingredients for life develop in the first place? I still could not believe that systems - heart and arteries and hemoglobin all working in cooperation together - developed step-by-step from nothing. I still didn't buy it.[7]

I didn't care whether or not there were four billion years of time for life to develop, I looked at the biological world and saw wrenches, wrenches everywhere in the "Evolution did it" explanation of origins. I didn't have Dr. Stillwell's kind of faith.

But (sigh) I did have spiritual superpowers...

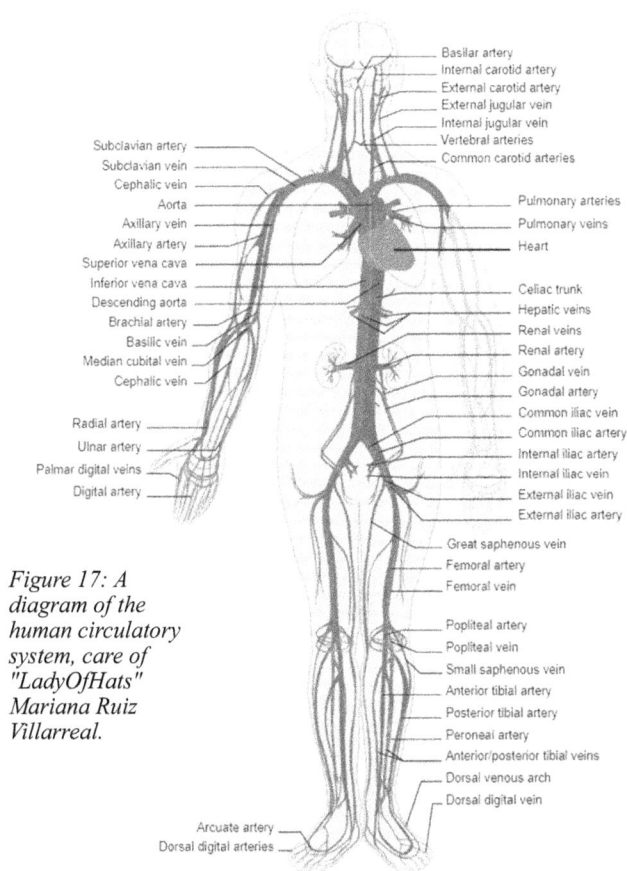

Figure 17: A diagram of the human circulatory system, care of "LadyOfHats" Mariana Ruiz Villarreal.

---

7 See "A Simple Light-Sensitive Spot" in the Appendix.

Summer 2011

Chapter 22

The Grand Canyon

Emily: What is this rock?
Dr. S: It's a breccia.
Flash: Breccia didn't know that.

And so the semester ended, that left-brained semester filled with chemistry classes. My dear friend Emily Gilmore and I rode in the van on our way to the Baltimore airport, checking our semester grades on her phone. The power of a touch screen with Internet access astonished me; it was the first time I'd ever handled a smartphone.

"I got all A's!" Emily rejoiced.

"Good job." I'd been humbled by a single B every semester so far. I checked my list of grades on Emily's phone, and I too had busted through with straight A's. Booyah! I must have scored high enough on the awful Organic Chemistry ACS test.

There were a multitude of reasons to rejoice. It was a gorgeous day in early May, school was out, and we were on our way to the Grand Canyon. What could be better than that!

In the end, eight students had handed over money for the trip: a comfy-sized group. Emily, Kyra, Katie G., Hannah and I made up the girl half of the pie, while Jared, Brennan, Don, Flash and Dr. S. made up the guy half. We flew out west (on one plane) and landed safely in the desert outside the Gambling Capital of the World. It didn't take us long to pile into rented mini-vans and head out to the grocery store.

"I'm not in charge," Flash insisted. "I'm just the bus driver."

We needed nine days' worth of supplies for ten people, and we quickly filled several shopping carts with food. Bagels. So many bagels. While Flash and the other students packed all our groceries into the vans, Dr. Stillwell grabbed me to trot back into the store for a few forgotten items. We were on our adventure!

So very conveniently, Mark Jones had moved near Las Vegas to serve as the principal of a public school there. Mark had served as Randy's best man at our wedding, and we were still friends. I called up Mark and asked if our small troupe of students and professors could crash at his house our first night after landing, and he'd said, "Sure! I'll make BBQ! You can all swim in the pool!"

Mark hosted our small tribe with warmth and generosity, and Dr. Stillwell ended up chatting with him all evening. It turned out they had attended the same university and knew the same people. They made a pair of quick buddies. Thank you, Mark!

Wednesday, May 11, 2011

Mark's House:

Jared:	Who is going to wake up the girls?
Dr. S:	Okay. I guess I'll be the evil one. [He leaves.]
Jared:	You wouldn't want him to change character.
AJ:	He does evil well?
Jared:	It just rolls off of him.

Driving through the Arizona Desert:

AJ:	Tell me about the geology here. Dictate to me.
Dr. S:	Dictate to you?
AJ:	I'm not going to say that a lot.
Dr. S:	I *know*.

Grand Canyon First Impressions:

The Girls: It doesn't look *real*. It looks like a giant backdrop.
Flash: Honestly, it's a little bit of a letdown. I really like the view. Aesthetically, it's colorful and grand, but it doesn't compare to other emotional reactions I've had. (You can't write that down, Amy Joy.)

45 minutes later...

Flash: Okay, second impression, Amy Joy. It's growing on me.
Dr. S: What you do is look down there toward the bottom of the Red Wall, and you look at the people down there, and you imagine yourself down there and looking up, and you get more perspective...
Flash: CURSE YOU FRACTAL CANYON AND YOUR FINE STRUCTURE!

Walking the Rim:

AJ: So, if I put on this sunscreen, will the bears want to eat me?
Flash: There aren't any bears out here. I'd have to see a lot of data on that. With no data points, we could have either no bears or else an infinite number of bears.
[We all laugh loudly.]
AJ: You can totally tell that we're a bunch of scientists - that we laugh at that.

6:15 p.m. Ten X Campsite
Chili and rice are cooking. The fire is going.

Flash: How do you start your journals, Don? "Dear Diary, Today I saw the darndest thing. Amy Joy fell over trying to catch a peanut."

(Laughter.)

Dr. S: Who had batteries with their dinner?
[The poor pack of batteries sat on the table. Innocent.]

Dr. S: You know, you'd have to use salt with them. But then it would be a salt and battery.

(Groans. So many groans.)

Figure 18: Our good geology professor on the edge of the South Rim of Grand Canyon.

I woke early on Friday to the sound of Dr. Stillwell on the phone with his wife. She was planning to fly out and join her husband for their own adventure late the next week, when she wouldn't be forced to sleep on the ground. In the meanwhile, he called her every morning about the time the birds ruffled awake. We were three hours behind the east coast, and 4:30 a.m. in Arizona seemed the geologist's favorite time to chat with his spouse.

Kyra still dozed in the sleeping bag beside me. I rested on my back, staring at the nylon ceiling of our tent and listening to the birds twittering away in the pine trees outside. By the end of the trip, I'd be squeezing in between Katie and Jared and Emily in one tent, all nice and warm and toasty like chipmunks in a nest. That early in the trip, Kyra and I were new to each other. I don't think our sleeping bags touched, and it was cold cold cold in our tent.

After awhile, Dr. Stillwell walked around and called for us to rise and make merry. We tumbled out of bed and grabbed apples and bagels to munch. We were venturing into the canyon that day, and we wanted to hike down and back up before it got too hot in the afternoon.

Figure 19: Jared and Emily, heroes extraordinaire, saving the author at the South Rim. Photo by Kyra W.

The problem with hiking down *into* the Grand Canyon is that one is then obliged to climb back *out*.

The prior day, we'd strolled the South Rim as an introduction. Kyra got some good photos of me "falling" off the edge and Jared and Emily "saving" me. We were pleased with our theatrics.

Dr. Stillwell shook his head. "Somebody's got to do it every year."

Today, though. Today would be a bit more arduous. Our descent started cheerfully enough. Flash began chanting out the canyon's formations from the top downward, trying to drill them into his head.

Flash:	Kaibab. Toroweap...
Dr. S:	Think of a sad bull. Toro weep.
Flash:	Haha... Kaibab. Toroweap. Coconino – not Grand Torino – Hermit Shale. Supai. Red Wall...

The Colorado River drifted along, a small blue ribbon from our view a mile above it. The cool morning air caressed our faces, and cactus flowers greeted us around every other turn along the hard rock path. As we walked along, people kept saying funny things, and I repeatedly pulled out my journal to write them down.

Katie G:	It's not a promise ring. It's not an engagement ring. It's a 'I want to be with you forever, but I can't get married yet' ring.
Dr. S:	It's a 'hedging-your-bets-until-something-else-better-comes-along' ring.
Dr. S:	I've been wanting to call Amanda and say, "Hello Amanda. I'm checking in to see how Jeff is doing. What? Trip out west? What are you talking about?"
Flash:	Yeah, except that I've been calling her from *your* phone. So, that won't work.

We'd talked about making the full 17-mile hike through the canyon, but that required reserved camp sites at the bottom. In the end, we decided the shorter hike would do. Beyond Skeleton Point, the trail dropped steeply into the canyon, and we were content with a mere three miles to climb out. We settled down at the point to enjoy lunch.

Heat had steadily increased around us as we dropped through the colorful sedimentary deposits. The sun had risen high, and the heat island effect of all that rock created an oven.

As we sat munching our bagel sandwiches, foolish people in flipflops walked past us.

"It's a heavy hike back out," Dr. Stillwell cautioned them, eyeing their tender footwear. "It's smart to have good boots if you're going any farther."

The flip-flop-wearers ignored our professor's pleasant warning. They just smiled at us and sauntered past.

We shrugged our shoulders, chewed on our sandwiches, and enjoyed the view. If our fellow hikers required a helicopter to haul them out later, we couldn't help that. We believed in freedom. They were free agents. They could pay their own fines.

Dr. S:	Okay. It's about time to go.
AJ:	Can I put up a fight?
Dr. S:	Yes.
AJ:	Can we stay for a few more minutes?
Dr. S:	No.
Flash:	It's a flexible itinerary with an iron clad time schedule.

Dr. Stillwell laughed out loud and declared that Flash had made the quote of the day. We finished our sandwiches at the edge of the cliff, brushed the crumbs off our laps and packed back up.

Figure 20: The trail we took down to Skeleton Point.

Finally, we sucked in a long last view of the canyon from Skeleton Point, then we turned and headed back up the red rock path. Emily and Dr. Stillwell tailed behind everybody else, and I joined them.

Emily liked taking things easy. Emily felt no obligation to hurry, because she wanted to relish her experience inside this particular gash in Earth's crust, and she enjoyed stopping to take pictures. She hung behind us most of the way back, filling her camera with digital delights.

On the other hand, Dr. Stillwell acted somewhat out of shape. He and I had led the pack to the top of Seneca Rocks the previous September, but the winter appeared to have replaced our professor's muscles with some adipose tissue. He decided to take his time too.

I hung back, grateful to Emily and Dr. S., because I didn't like getting hot. I'm pale and Nordic, and I heatstroke easily. If I do too much work in the hot sun, my head feels like somebody stuck a bicycle pump in one ear and overinflated my skull. It's the downside to those mutations that make me well-suited for northern climates. While the other hikers disappeared out of sight ahead of us, we three strolled leisurely.

We had to trek single file along the path, so we switched back and forth, taking turns walking in front. As we progressed in our slow march, Dr. Stillwell asked me to tell him a story.

"Tell me a story, Amy Joy."

So, I did. I told a couple of stories.

Figure 21: The hot, red trail up from Skeleton Point. Photo by Jared T.

Chapter 23

Fear

We have all known fear. Throat tightening, maddening fear. We've all jerked awake from childhood nightmares, hearts pounding, acid in our stomachs. We've pulled the covers over our faces and buried under our pillows to protect ourselves from the beasts that lurked beyond the edges of our sheets. We've walked past dark alleys or through black woods, where unknown dangers tormented our imaginations.

I was 18. It was just two months into my first college career as a philosophy major. I'd been dropped off at the Skate King in Bellevue, Washington, the happy little skating rink where I'd spent many hours as a child. Friends were supposed to meet me there at 6:30 p.m., but I had waited and waited and waited and waited, and they never showed up. Nobody answered in their dorm rooms when I tried a pay phone, back in those days before cell phones were common.

Finally, about 8:00, I decided I'd walk back to the dorms.

It was a decent little stroll. Maybe four miles, but I only had to hike through Bridle Trails, the neighborhood of expensive horse owners, upscale properties where people drove BMWs and Jaguars. I had no reason to be afraid. The previous year, I'd regularly walked three miles to school through industrial Spokane, where the hobos warned me against jumping trains and hurting myself. I'd enjoyed those walks. I'd enjoyed my talks with the hobos. The wealthy world of Bellevue, Washington should have been a piece of chocolate cream pie after industrial Spokane.

I don't know what filled me with terror as I left the Skate King parking lot under a clear October night sky. I don't know why fear overwhelmed me as I plodded up that first hill, towered over by familiar fir trees. I had risen at 2am to go deer watching when I was 12. I had walked through the dark basement of a haunted house without worry. I wasn't afraid of the dark. That singular night, terror overwhelmed me as I walked up the peaceful street. I panicked that I'd get attacked from the bushes and murdered, right there on a Bellevue sidewalk.

As I marched, I wrestled against the fear, struggling unsuccessfully to throw it off. It clung to me like a cold, wet blanket. "Please don't let me die, Lord," I tried to pray. "Please don't let me die. Please don't let me die." I didn't feel any better. "If You want me to die, I'll die, but I'm going to go out *fighting*." It was awful. I couldn't shake it. Every bush held a crouching enemy waiting to grab me.

I hiked several more yards, and soon the wide, dark parking lot of Westminster Chapel opened up beside me. Maybe I could find refuge there. I searched across the vast, empty lot. No. It didn't look hopeful.

"Don't let me die," I begged God. "Don't let me die."

Then, I heard God in my head for the first and only time in my life. God had spoken to my spirit before, but this time I actually heard Him speak into my head.

"Peace, be still."

And instantly, I was.

Those three words were a verse, and I knew the whole thing, the whole context of the passage. Those were the words Jesus used during the storm in Mark 4. Jesus had been sleeping in a boat while the disciples traveled across the lake. The Bible doesn't tell us about Jesus' exhaustion, but I imagine it took a lot of energy to heal people all day long. I can imagine weariness hung on Him like a cold, wet blanket, dragging Him down until He eased into the bottom of a fishing boat and fell asleep. A fierce storm hit, and even that didn't faze Him. Finally, the disciples shook Jesus saying, "Wake up! Get up! Don't you care we're all about to die!" At their fear, Jesus struggled up. He stood as the boat pitched and dove in those wild waters and solidly told the wind and the waves, "Peace be still." Instantly, everything went calm.

That's what happened inside of me. A storm of fear plagued me. The Lord told my heart, "Peace, be still," and the storm inside me vanished. Snap. Just like that. Not only did the storm end, but the terror was *replaced*. That terror disappeared in an instant, and it was replaced by an utterly comforting confidence.

It was an amazing experience. I suddenly comprehended deep inside me that it didn't matter whether I was in the jungles of Vietnam or at home in my bed, God was protecting me. He kept me safe, and I didn't have to worry. Psalm 91 had come to life in my heart:

> *A thousand may fall at your side, And ten thousand at your right hand; But it shall not come near you. Only with your eyes shall you look, And see the reward of the wicked. Because you have made the LORD, who is my refuge, Even the Most High, your dwelling place, No evil shall befall you, Nor shall any plague come near your dwelling; For He shall give His angels charge over you, To keep you in all your ways. In their hands they shall bear you up, Lest you dash your foot against a stone.*
>
> Psalm 91:7-12

I walked those four miles back to my dorm in complete security.

A month later, Brian Reed's car rolled into the median at 55 miles-per-hour, and I bounced around in the back seat without a seatbelt on. I emerged from that smashed, destroyed vehicle completely unharmed. Hillary up front said she saw angels flying around and around me, keeping me safe.

Sometimes I still feel fear. I'm not immune to it. I still get dizzy when I'm on tall buildings or bluffs overlooking the ocean. Heights and other people's driving scare me. I know that I shouldn't be afraid, though, because I'm protected. I've always been protected.

Dr. Stillwell wanted me to tell a story. So, I told him about that night. I reminded him that I didn't normally have a problem with fear, that I tromped around in the dark in the middle of the night, no worries, but this particular night I couldn't shake my terror.

"And I was fighting it, feeling ridiculous. I felt afraid I was going to die, and I didn't want to die. If God was going to let me die, then I'd die, but I wasn't going to go down without a fight."

Dr. Stillwell chuckled behind me.

"So, I was climbing up a hill. You know how Bellevue is - all hills. And I was trying to pray, because I was so scared. It wasn't Detroit or Chicago or even Seattle. It was calm little, suburban Bellevue where everybody drives a Mercedes or BMW. And as I was walking up the hill, I heard God speak into my head for the first time in my life. He simply said, 'Peace, be still.'"

I glanced back at Dr. Stillwell on the path behind me. "It's like when Jesus told the storm to stop on the Sea of Galilee. Jesus told the wind and rain, 'Peace, be still,' and the storm instantly died. That's what happened to me. I was terrified one second, with a storm raging inside of me, thinking it didn't even matter to God if I died. Then, He spoke, and instantly my soul calmed. I didn't just stop feeling scared; my fear was replaced by a confidence, by this massive confidence that I was going to be okay, that I was safe, no matter where I was. If I were in the jungles of Vietnam, it didn't matter. I'd be just as safe as if I were in my bed at home."

Dr. Stillwell freely allowed me to tell him these tales from my life. He listened to me, and that always impressed me. He disagreed with my interpretation of events, but he listened quietly, and he didn't attack me or laugh.

We marched up out of the canyon, Dr. Stillwell, Emily, and I. We hiked up those switchbacks the three miles up to the rim, and I told other stories. We took our time and enjoyed our conversations. We found the rest of our group talking with a park ranger, and all of us climbed the final mile together.

Then. Then, Dr. Stillwell treated us all to showers and pizza. Wonderful pizza.

Two days later, Dr. Stillwell said quietly, "You know… when you were telling your story about being scared that night in Bellevue, I wanted to say something about Jeffrey Dahmer."

"Yeah, or Ted Bundy," I agreed.

Dr. Snyder knew Bellevue. He had grown up in that area too, and I had a right to be afraid, I suppose. Serial killers were a part of the deceptively pleasant Bellevue world of three car garages and ballet lessons. The high number of serial killers in the Northwest is noteworthy. We didn't have just Ted Bundy. The Green River Killer, the East Side Killer, the I-5 Killer were all men who haunted my childhood.

In fact, I think we had a run-in with a highway slayer when I was six-years-old. I'll tell that story too.

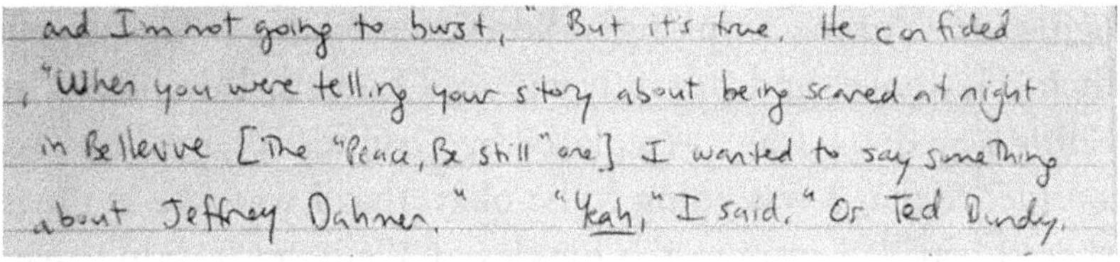

Figure 22: Excerpt from the author's journal, May 16, 2011.

Chapter 24

Angry Man at the Window

My mother always ran out of gas on the freeway. It was her thing. I don't think she had an aversion to buying gas, but maybe the gauges didn't work anymore in that ancient blue Oldsmobile Grandpa passed down to Dad. I think that was part of it, and I also think Mom was always low on cash. Occasions when we coasted into a gas station were times for great rejoicing.

In those days before car phones, Mom took advantage of anybody who would help us. An outgoing, fearless sort of person, she would get out and flag down passing cars. She'd pack us into a stranger's vehicle for a ride to the gas station.

This had happened.

I think my mother is not like the mothers of most children. Perhaps if she were more fearful, we'd have run out of gas less often.

On this occasion, Mom had collected me after school from first grade, and we drove down the freeway from Issaquah to Bellevue. Timmy Holmes sat in the back of the ancient Olds with me, because Mom was watching him that afternoon. Timmy was about eight-years-old. We drove down I-90, passing Lake Sammamish State Park where Ted Bundy had killed at least two young ladies in the 1970s. We swam at Lake Sammamish all the time during the summer, and we'd heard the stories about how Bundy pretended to have a broken arm so he could lure women into his Volkswagen Beetle.

As we passed the lake on the freeway, the car engine started catching, making the familiar death coughs of the empty gas tank. My mom pulled over to the side of the road and said, "We'll just wait for a policeman." We sat there for about five minutes, waiting.

A state patrolman passed us going the wrong way, and Mom said, "He'll turn around and come back. It will be just a few minutes." So, we waited. And waited.

After a bit, a pickup truck pulled in behind us. My mother usually got out of the car to talk to people. This time she did not. She looked in her rearview mirror and ordered, "Kids. Lock the doors."

I reached up and shoved on the heavy lock beside my head. It was solid and rounded under my palm, and I pushed until I felt it thunk. On the other side of the car, Timmy's lock thunked too.

"That was a miracle," Mom will say to this day. "You guys never did anything just because I told you to."

Well. She'd sounded like she meant it.

A moment later, a man in a jean jacket walked past me and stopped at the driver's window. Mom rolled it down the smallest bit and spoke to him through the crack.

"Hi," he said nicely. "You guys okay? You need any help?"

"Thanks," my mother said. "We'll be fine. We're just waiting for the highway patrol."

Thinking back, I realize my mother was a pretty woman, one of those original California girl types.

I couldn't see the face of the man. He leaned over to speak at the crack in the window, but I only saw his chest. He wasn't particularly tall.

"It's okay," the man said. "It's no big deal. I'll be glad to give you a ride."

"No," Mom stood her ground. "We'll be fine. I can't go with you and leave the kids here, for obvious reasons. And I can't take the kids along, for obvious reasons."

The guy didn't give up. He didn't say, "Suit yourself," like a normal person.

"No, it's really no big deal," he insisted. "You don't have to sit here beside the road with kids in the car. You can all fit."

I understood that there was no way on God's green earth my mother was going with this person. There wasn't even a question in my mind. He should have figured it out, because she didn't get out of the car when he pulled over. She'd told us to lock the doors. She only cracked the window for him. She wasn't going to open the door to him. He should give up and go away.

But, he didn't.

He got mad. And his voice rose. "I'm already pulled over. There's room in my truck. I'll just take you to a gas station."

"You go ahead and go. If you go call the police for us and let them know we're here, that would be great."

It took time for him to give up. He kept trying to get Mom to change her mind. Finally, still angry, he marched back past my window.

Using her side mirror, Mom watched him walk back to his pickup. As he stepped up into the cab, the bottom of his jean jacket caught on the handle of a large knife he'd stuck into the back of his pants. It wasn't a knife in a sheath on his belt, like a lot of guys have. It was tucked into the small of his back, hidden by his jean jacket.

I've since wondered if he was Randall Woodfield, the I-5 killer, but the dates don't match up. Woodfield was suspected in about 44 murders and 60 sexual assaults up and down the I-5 corridor from Washington to California, but he was caught before I started the first grade. "The Green River Killer" Gary Ridgeway's murders fit the timeline, but he strangled his victims. So, I don't know. We had a lot of serial killers in the 1980s.

The event was memorable for a couple of reasons. First, it took the guy a long time to give up, even after my mother repeatedly told him, "No." Second, the news later reported that a woman was assaulted and stabbed along that same stretch of freeway. Mom made a big deal about it, because it was in the news the next day or so.

All I do know is that my mother meant it when she told us to lock our doors.

After the guy drove off in his truck, Mom used the rearview mirror to look at us in the back seat. "Did you guys hear it? Did you hear the voice?"

"What?" I asked.

"When he pulled up, did you hear the voice?" Mom asked again.

"A voice told me, '*Do not* go with this man.'"

    We didn't hear it.

    "You mean you just obeyed me?" she said.

    Yep. We'd just obeyed her.

Chapter 25
Zion

Flash: I just think it's amazing that all those wars were fought at national parks!
AJ: I know! What a weird coincidence, huh?
Flash: It's just crazy!
Dr. S: It's because it's public property. That's why they fought there.
Flash: That makes a good argument for private property.
Dr. S: If it were private, they wouldn't fight any more wars.

In the days of the kings of Israel, Mount Zion was the hilltop where the Temple gazed out over Jerusalem. Mount Zion, the sunny hill. Zion is a place of great historical and prophetic significance.

Zion is also the name of a fantastic national park in southern Utah. It's a gorgeous spot where rich red cliffs rise up and up to white summits, where wind-swept crossbedding swoops back and forth, forming distinct zig-zag designs. I love the staircase of formations in the Colorado Plateau, and I think it's fun that God chose to give me a prophecy while our geology troupe visited the red cliffs of Zion National Park. It seems fitting that God chose that spot to give me a hint about the future of Dr. Paul Denali Stillwell.

In 12 pages, I'll tell you about this prophecy, but I'm taking a risk in doing it. The picture God showed me at Zion is still future as I write these words, and I hope it comes to pass sooner than later. It does involve frustration and pain, but it's got a happy ending. Also,

I don't want to be stoned as a false prophet.[1]

Not that Zion started *off* full of power and glory for me. And it didn't end with power and glory for me either. Bagels. That's how it ended for me. Eating bagels. There were some chewy bits in the middle, though, and those gave it a little kick.

Our first full day in Zion National Park was Monday, and we decided to make the four-mile hike up to Observation Point. We were excited about the fact that we'd do all our hard up-hill climbing during the morning and hike down at the end – the precise opposite of our trek at the Grand Canyon.

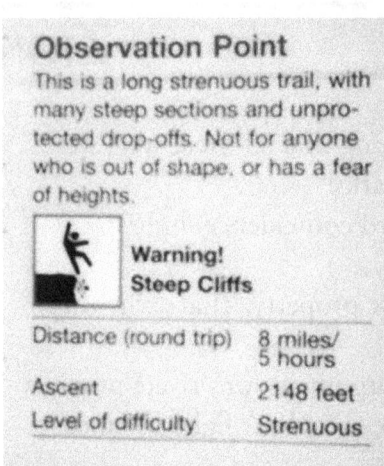

There's so much to see in Zion! Crossbedding! Great views! Cool rock formations carved by wind and floods. We had a great time joking together up the switchbacks through layers of rock, admiring vast views and checking out ravines and little caves.

"Hmmm..." I peered into a narrow but deep crevice. "I could lean across and catch myself against the wall there with my hands."

"If you do that, Amy Joy, you will be stuck," Dr. Stillwell said dryly. "And then you'll end up leaning over that hole all night. We might leave you a cookie," he chuckled.

I partially believed him.

Dr. Stillwell seemed to have gotten his hiking legs back, and soon he and Hannah and Brennan marched up ahead of everybody. Some of the girls and I hung back and took our time.

After awhile, I ended up on my own. I admired the windswept sandstone walls as I hiked. I tried to imagine the processes that formed their swooping, back-and-forth pattern in the creamy Navajo Sandstone. I admired the work of the park service in carving a path through the rock for us.

---

1   Deuteronomy 18:20-22

Figure 23: The view from Observation Point in Zion National Park. Those fuzzy things at the bottom are full grown trees.

After awhile, I got concerned that I'd fallen behind, so I started hiking faster. The path continued on and on, back and forth, and I'd finish one set of switchbacks only to find another. Where was everybody? I stalwartly pushed myself onward, but I wasn't catching up to the people ahead.

Then, I turned a corner, and a most beautiful view spread out below me. The whole of the canyon stretched out in a green cascade through red cliffs. It was gorgeous and open, a postcard in real life with the Virgin River winding along at the bottom. I expected flags of a spired castle to wave at me from across the jagged valley.

Then. Then I remembered standing down at the river that morning, looking up. I remembered that what I saw from the river was a sheer cliff face up to the top, a cliff face thousands of feet into the air. I was now standing near the top, at the edge of that wall of red rock.

The Grand Canyon drops a mile into the earth, but those cliffs weren't scary. They had levels. I could see the next tier below me at Grand Canyon. This narrow path was cut into the side of a cruel flat wall. Also, I'd had my friends around me at Grand Canyon. Here, I stood isolated on the edge of an abyss. As soon as I realized my position, I jumped to the wall a foot behind me and gripped its gritty surface with both hands. My breaths caught in my throat. I had to force myself to inhale and exhale. Inhale. Exhale. It's a bad thing, a fear of heights, because it makes one dizzy and weak. I had to stop and rest a moment.

Oh God. Oh God. Please don't let me die.

A few feet from me loomed a 2000-foot drop of death, and the wind blustered at me, slapping bits of dirt and rock against my cheeks. A particularly brutal gust slammed against me, and I imagined the wind could gust enough to toss me off. Feeling the real danger of my situation, life-and-death questions jumped through my mind.

If I fell, would God save my life and let me float down and waft into the river? Would He slow my fall, or would He (wind blowing whoooooooooohhhhh) let me crash?

If I died, would He give me the option to come back to life or leave me to sleep until Judgment Day?

"I can't die!" I thought. "I have little children! And Dr. Stillwell won't take students on hikes ever again!"

I hugged the cliff behind me and edged carefully along the ledge, imagining the different scenarios if I fell. I didn't want to bang against the cliff repeatedly as I tumbled to my death. Breathe in. Breathe out. I fought the urge to crawl on my hands and knees as wind banged against me on that exposed rock face, spraying sand into my eyes. No, I'd resist. I'd stay on my feet! I tucked my head down and let my straw hat catch most of the attack, the gusts rushing and whistling past. Then whoosh! One blast tore the hat from my head and flung it off the edge.

Figure 24: The narrow path up to Observation Point in Zion National Park.

That wasn't very funny. That wasn't funny at all.

I had just finished telling Dr. Stillwell that God had promised to keep me safe wherever I was. I had just told him that I was as safe in the jungle as I was in my own bed, and here I was scared because of a silly little 2000-foot cliff? What was wrong with me? Didn't I believe what I'd told him? Yes! But, I didn't *feel* safe!

Holding the sturdy wall behind me, I kept edging my way up the path. "I'm safe," I told myself. "I'm safe. I'm safe. I'm safe. I'm safe." Oh God. Oh God help me. "I hate heights. I hate heights! But I'm safe."

After the first switchback, it was easier. There was a buffer offered by the tiering effect, but it didn't soothe me much. If I tumbled down to the lower ledge, I'd still bounce into eternity.

The wind blustered and wooshed as I climbed those few switchbacks on the cliff face. Then, I reached a little tree, a brave little pine tree clinging to the edge, way up there at the top of the world. I loved that little tree. I settled into its shade and enjoyed the fact that it blocked the wind. I felt like Jonah under his gourd plant overlooking Nineveh, and I decided to sit there for awhile, just sit there and rest beside my tree of refuge.

Two or three minute later, good old Jared appeared around the corner. Oh wow. I didn't know Jared was behind me. I thought we girls were the last ones.

I looked at him for a moment and he looked at me. "AhhhhhhhhH!!" I shouted at him. "Ahhhhhhhhh!!!"

He grinned and joined me with my new best-friend-the-tree. "I know, right?"

I don't think he really felt that "ahhh!" in his heart like I did.

"The wind took my hat." My hair was a mess, sweaty and wind-styled, with sand all through it.

"I'll go get it," Jared said.

"No!"

"It's all good. It's down there? I'll get it."

"No, Jared. No!" I didn't care that he was joking. "It's gone! Let it go!"

Soon Katie arrived, and together we trekked the rest of the way to the top. In a few minutes, the path opened into a copse of scrub evergreen trees that crowned that 2148-foot chunk of red rock. It felt better to cheat death with friends beside me, but I still found the wooded area too small. The vertigo lingered, and I wound my way between the trunks, feeling like the whole cliff top could slide off into oblivion.

We found Dr. Stillwell and Brennan and Hannah resting happily at the outlook just ahead of us. They were it. No wonder I didn't catch up with the pack; everybody else had been behind me after all! I marched up to Dr. Stillwell and said, "Wow, you guys took off!"

"Hannah trained all spring on her Stairmaster," Dr. S. said.

"Brennan and I were just trying to keep up!"

I looked beyond him to the vast view that gave the trail its name of Observation Point. The valley spread out below. "I need a hug," I said to him. "Can I give you a hug?"

"Okay," he said.

That's when I gave Dr. Stillwell an actual factual hug for the first time in our history. I really needed it, and I was grateful he didn't say, "Um. No."

Then, I sat down and eased closer to the edge and let the blood pump oxygen into my brain. A chipmunk hopped back and forth over the rocks, hoping we'd brought him a treat. We sat and watched a peregrine falcon circle overhead. I edged slowly to the brink, sliding my feet little by little over the sandy rocks. The river meandered like a brown inchworm in the canyon below. An open land greeted me, wild and fresh and alive. It was a good view.

The rest of our group wandered through the trees, and we sat around chatting out our emotions. I pulled my journal out of my backpack and had everybody write their impressions. I even got one Youssef Saliba from southern California to write a note. He said, "I made it to the top, and I am most definitely alive!!" Exactly, Youssef. Exactly.

We spent a solid bit of time up there, enjoying the visual fruits of our labor, but eventually it was time to leave.

Dr. S:	Are you ready? We're going to start easing down.
AJ:	Okay.
Dr. S:	Are you sure? I don't want to push you.
AJ:	Then you could see me fly.
Dr. S:	Oh, I wouldn't do that. Although, I do want to know, 'Does she bounce? Doesn't she bounce?' … It's not one of the most *burning* questions I have.

Chapter 26

Sunshine and Prophecy

Dr. Dan: What's free induction decay?
AJ: It's what's happening to my teeth.

Dr. Stillwell scolded me several times during those nine days in the desert. He scolded me for not keeping up with the group, and he scolded me for calling Dr. Gurden, "Flash." He threatened to strap down my arms because I wrote in my journal too much, and he told me to be good when I wanted to toss snowballs at Jared at the top of Bryce Canyon.

C'mon Dr. Stillwell. You should have thrown snowballs at Jared too.

Flash captured a video of me chucking a snowball down through the natural arch at the top of Bryce. It's all slow motion, and the snowball flies out across the rock-strewn hillside all the way down through the arch.

Dr. Stillwell didn't throw a single snowball. Silly man.

Emily pointed out that Dr. Stillwell would herd us along at times, saying, "Come… come…" She quietly mocked him in a Yoda voice. Flash didn't *say* anything in response to Emily's light-hearted jeers, but he smiled every time.

"Be good, Amy Joy," Dr. Stillwell told me.

I didn't want to be good! I wanted a snowball fight!

That particular hike down from Observation Point in Zion, the geologist chided me for my clothing choices.

"Gosh, my hands are swollen," I noted as we trudged back through the switchbacks.

"How can you tell your hands are swollen?" Flash asked.

"Because I've been pregnant three times."

"It would help if you wore shorts," Dr. Stillwell griped at me after a few minutes. "It's too hot here for you to wear jeans."

"But jeans protect my legs," I explained. "I get banged up a lot."

A few minutes later, Dr. Stillwell said something unexpected. We rounded a corner on the last set of switchbacks, and he said, "You hide things."

I hide things?

"You're an open person and you seem very willing to speak your mind," he said, and I nodded in agreement. "But I think it's a front to keep attention off other things that you don't want to share. You chicken out."

I wondered what brought that on. Was it my jeans? Was he making my jeans a metaphor for self-protection? I guessed that he'd been processing our conversations over the previous week, the previous year, trying to decide how to interpret the variety of things I'd told him. He contemplated that I might be far more emotionally and psychologically damaged than I let on. *That* was his personal explanation for some of the things I'd said? Did ... did he think I was traumatized and slightly crazy? Is that what he thought?

He misunderstood me. The reason that I appeared open and exposed to the world was simply because I truly wanted people to see inside me. I didn't mind being judged as much as I hated being *mis*judged, and I didn't have the energy to be anything other than myself.

Obviously, I didn't tell Dr. Stillwell every dark horror, but that was only because I made efforts toward decency. "I used to tell people more," I explained as we trekked down the final stretches of red rock. "But, then I realized it wasn't good to do that to them. If I don't share things, it's not because I'm a chicken. If I told people all my problems, I'd be putting my burdens on their shoulders, and that's not kind. So, I stopped telling people everything, because I didn't need them to bear my burdens."

Dr. Stillwell nodded at that.

What... what did he think I was hiding anyway? I didn't feel like I was hiding anything.

When we finally reached the bottom, the good doctor and I ended up sitting across from each other on the shuttle bus back to our campground. I pulled out my journal and began scrawling out a note to him. I wrote:

> Dr. Stillwell. There's a lot to it, but mostly, I work very hard to not do anything that I don't want to admit or be honest about. Remember that day I fibbed to Dr. Zenith about the percent errors? I had to go and tell him not to give me credit for them because I hadn't had them done. I don't like having to do that. So I work very hard not to do things I'd not like to admit to doing.
>
> Everything gets found out somehow. It always does. So, you might as well be up front about your weaknesses, and that doesn't bother anybody. They would rather know up front rather than being unpleasantly surprised. And besides, who likes somebody who pretends they don't have anything wrong with them?

I handed him my journal, and he read what I wrote. He agreed with the first paragraph. "I operate on a similar principle," he said. Then he skimmed down further and shook his head.

"No," he said. "I wasn't referring to matters of your character or your weaknesses. I mean, you have a lot of pain. And that's what you hide."

"I'm not a chicken about it, though. I don't think I hide it at all." Then I grinned. "But, you can think I'm awesome if you want to."

I expected Dr. Stillwell to roll his eyes. I expected him to say, "Whatever." But, he didn't.

"I know that's the sun," he pointed out the window of the bus, where the sun beamed cheerily above the canyon cliffs. "I know that's the sun. I don't think it is. I know it is. I don't *think* you are. I know it."

That caught me in the chest. I was prepared for Dr. S. to be cross with me because I wore jeans. I was prepared for him to call me a chicken. I'm okay with criticisms and rudeness and attacks, which is why I can expose myself to the world. However, I wasn't prepared for his free words of esteem, offered without explanation or apology. He spoke the words simply and plainly, and they were like the light from the sun shining into the bus. I have treasured that moment in the years since. I think back on it at times when he's frustrated with me, and I wish I could go back to those few minutes together when he dropped all his grouchy professor ways and just offered me straight love.

I'd worried about Dr. Stillwell during Spring Break, and God had answered my concerns by letting me know He cared about the geologist. I believed Him. I did. I knew God had His hand on the codger. Still, I started to worry about Dr. Stillwell again that day in Zion.

I honestly think that Dr. Stillwell is in a good position to understand God. He wants me to trust his judgment, even if I don't understand everything he's got going on. He has all these students that he's responsible for, and he gets irritated when they do things potentially harmful to themselves and others. He was willing to let me experience the consequences of my own foolishness if I fell over the narrow cavern and had to hold myself there all night. Cookie offers notwithstanding.

Actually... actually... I take that back. God is a lot more patient and compassionate than Dr. Stillwell.

Yet, when I told Dr. Stillwell stories about things God had done in my life, he resisted my personal narrative, my experiences. He didn't take me at face-value. Instead, Dr. Stillwell started pondering whether I was psychologically damaged and traumatized. He didn't want to believe that I was telling him something real, something true.

It was on our day trip to Bryce Canyon that Dr. Stillwell gave me as open an answer to his God problem as ever I'd heard. It started because I told him about my herniated belly button and how I'd been healed as a baby. "I don't know how to deal with that scientifically," I told Dr. Stillwell in the driver's seat beside me. "But, it happened."

"It's part of your family's oral tradition. I don't have enough data to make an interpretation of the event."

I didn't let up. "My grandmother called my mother to tell her to look at my belly button. She called from two counties away. She called my mother to tell her about the miracle, and not the other way around. There is a real explanation for that. Something real happened there."

My chest burned like crazy as we drove through the winding Utah canyonlands. It burned away, and I tried to explain to Dr. Stillwell about it – about that burning in my chest. It used to only burn when I was worshiping, but it had become far more constant during the 20 months since Randy had died.

While the professor and I started down the path of a long discussion, Jared and Katie and Hannah sat in the back, earphones on, dozing as far as I could tell.

Finally, Dr. Stillwell spoke his heart to me. Not completely. He's the one who hides his pain, who diverts attention from the old wounds in his soul. He's amazing at projection, Dr. Stillwell is. A true expert. Still, he opened up a crack that day to give me a taste of the bitterness he harbored so carefully. God hasn't ever shown me where it festered from, but I cared about my dear geology professor, and his pain mattered to me.

"You know what I think?" Dr. Stillwell said with emotion. "I think that if there is an all-knowing God who can intervene in human lives, if there is one, then He is entirely capricious and really cruel. And I don't want to have anything to do with such a deity."

He kept it brief, but he'd done it. He'd dropped all the show, the façade of politeness, and he released some of the unfettered harshness in his heart. And I was glad. *That* is what I'd always wanted from him.

I didn't argue with him. Why would I argue with him? He'd

showed me something precious; he'd stopped hiding for a moment.

I knew that God intended to have Dr. Stillwell with Him one day, that my crusty professor was going to know God himself, and the King of the Universe was going to show him His glory. Dr. Stillwell wasn't a lost cause, and his years-fired hardness were not beyond God's reach. As my heart burned within me, I felt only warmth for this man who'd just told me that he thought God was cruel and capricious.

But, fear had overwhelmed me up on that cliff, even though God had told me He'd keep me safe, and I was afraid down in the valley again, even though God had told me He had good plans for Dr. Stillwell. After our tram ride in Zion, I sat worrying about my friend. I worried that he was separated from God and in danger of being lost forever. "Lord," I prayed. "Please forgive him. Please."

God is so patient with me. He doesn't seem to get disgusted even when I feel I deserve it. He didn't turn away from me that day saying, "I already told you!" Instead, my Heavenly Father gave me a passage, and my heart warmed inside me as I read it. The passage is a prophecy, and since that day in Zion I've been waiting for its events to unfold. The prophecy is Job 33. In the context of the book, Elihu is defending God to Job. But, here's the picture it paints; it talks about a man who *thinks* God considers him an enemy. Elihu says:

Look, in this you are not righteous. I will answer you,
For God is greater than man. Why do you contend with Him?
For He does not give an accounting of any of His words.
For God may speak in one way, or in another, Yet man does
not perceive it. In a dream, in a vision of the night,
When deep sleep falls upon men, While slumbering on
their beds, Then He opens the ears of men, And seals their
instruction. In order to turn man from his deed, And conceal
pride from man...

Job 33:12-17

That was interesting all by itself. By the end of 2015, God would give me a total of four solid dreams about Dr. Stillwell. All four dreams explained to me what was going on with the professor - in advance. They prepared me, so that I saw through him when he behaved in ways that could have been misunderstood. Those dreams protected me, but they also protected my professor by preventing me from thinking the worst of him when the worst wasn't true.

I've never told him this, but I've asked God to give Dr. Stillwell dreams as well. I figured God could do that. I've never asked Dr. S. about it, and of course he's never shared any with me.

As far as I was concerned, that wasn't the significant part of the prophecy anyway. Those few verses were just a preface, and the important part was what happened next in Job 33. Elihu continues:

> *He keeps back his soul from the Pit, And his life from perishing by the sword. Man is also chastened with pain on his bed, And with strong pain in many of his bones, So that his life abhors bread, And his soul succulent food. His flesh wastes away from sight, And his bones stick out which once were not seen. Yes, his soul draws near the Pit, And his life to the executioners.*
>
> *If there is a messenger for him, A mediator, one among a thousand, To show man His uprightness, Then He is gracious to him, and says, 'Deliver him from going down to the Pit; I have found a ransom';*
>
> *His flesh shall be young like a child's, He shall return to the days of his youth. He shall pray to God, and He will delight in him, He shall see His face with joy, For He restores to man His righteousness. Then he looks at men and says, 'I have sinned, and perverted what was right, And it did not profit me.' He will redeem his soul from going down to the Pit, And his life shall see the light.*
>
> *"Behold, God works all these things, Twice, in fact, three*

> *times with a man, To bring back his soul from the Pit, That he may be enlightened with the light of life.*
>
> Job 33:18-30

I recognized that Elihu was talking to Job, and this was historical poetry. I recognized that I needed to be careful about jumping to conclusions. It's unwise to jump to conclusions. I also recognized there were parallels between me and Elihu. He was a young person speaking to elders whose wisdom he should have respected, but just because there were parallels… that didn't mean this passage was for me.

However, I had been learning to trust the burning in my chest and what it communicated to me. God had been using the Bible to talk to me about Dr. Stillwell for more than a year, and I've believed since that trip to Zion that Job 33 is a picture of Dr. Stillwell's future.

It's not a pretty story. It's not pretty at all. It indicates that he's going to be in horrible pain – so much that he won't eat. He'll lose weight until his bones stick out, and he'll lie awake at night, unable to sleep because of the pain in his bones.

Then, a messenger comes along, an interpreter, one among a thousand. The dying man is healed, and his youth and health are returned to him. He prays and God shows him favor, and he sees God's face with joy.

I hope the mediator is me, but it could be anybody. I've never ever prayed for somebody and seen them physically healed in front of me. I've told Dr. Stillwell several times, "One day when you're dying, would you please let me know?" The reality is that nobody in his family will tell me when he's dying unless he asks for me specifically. Either way, I have enough confidence in this picture that I'm daring to tell you - the whole world - right now, before it's taken place. I want it in print before it comes to pass. That's what we do. We make predictions, and if our predictions come true, we think we're onto something.

Otherwise… I'm just a fool. We'll find out, won't we?

I'm still waiting. I want God to heal Dr. Stillwell – to heal his body and to heal the mangled pieces of his heart. I want him to delight in God and God to delight in him – for them to deeply enjoy one another. I want Dr. S. restored to his youth, with time to enjoy life all fresh and new. I want that for him.

As I write this, I've never mentioned any of this to Dr. Stillwell. I've wanted to. I've written him emails that I've never sent, because it's always felt wrong. The time hasn't been right. More than four years have passed since our trip to Zion, and when I saw him two weeks ago, Dr. Stillwell's heart remained hard toward God. He's not in bed, dying of anything. His bones aren't sticking out. Actually, he's looking pretty healthy.

If he simply gets hit by a car one day… well… then I guess you all have permission to stone me.

Don't worry. I'll let you know.

Chapter 27

Stinky Bugs Reprised

Stink bugs. I discovered about 12 dead ones behind the face plate of one of the electrical outlets when I repainted the cabin kitchen. Get out the vacuum.

Olly olly oxen free!!

The brown marmorated stink bug is a remarkable creature. I don't see any good reason to keep it on the planet, but I've grown to admire its character.

Apparently, it smells like cilantro. I don't want it in my salsa, but I wasn't wrong to consider it a "green" smell.

The stink bugs are tough little guys who courageously resist death. They have durable brown shields on their backs to protect and camouflage them, and not a whole lot of creatures like to eat them. They gather together in large groups without fighting and have a certain attitude of the-more-the-merrier. When they find a good place to spend the winter, they send out a pheromone that says, "Join me!" Those little pests take care of each other.

They aren't good to us, but they are good to each other. They thrive, and I think we thrive too when we behave a little more like the brown marmorated stinkbug.

I have this idea that all creatures on the planet can teach us something - and that's one of the many reasons they're here:

> *The ants are a people not strong, yet they prepare their meat in the summer;*
>
> *The conies are but a feeble folk, yet make they their houses in the rocks;*
>
> *The locusts have no king, yet go they forth all of them by bands;*
>
> *The spider taketh hold with her hands, and is in kings' palaces.*
>
> Proverbs 30:25-28

What's the point? These critters all have weaknesses, but they find ways to overcome. Thus, we find them in abundance. A weakness is no reason to give up and die.

God is big into picture stories. He's into metaphors that help us visualize our own situation. Jesus spoke in parables a lot, but similes can be found throughout the Bible, and I think they're scattered throughout the animal kingdom as well.

Consider the female deep sea anglerfish. This swimming gargoyle is about the ugliest thing on the planet. She's a hag. Other anglerfish can be colorful or wield camouflage powers, and each attracts its prey by dangling a specialized fishing lure in front of its face. Most anglerfish are kinda cool. The deep sea anglerfish, though; that chick is the stuff of nightmares. She's got long, ice pick-like teeth. Rows of them. Her face is raw horror. She might very well be a devolved monster, a genetic mutant that has survived only because the deep sea hides her grisly visage from the eyes of her mate. I don't know.

The male anglerfish is a small thing that mates by fusing with the female. He loses his head and face - his whole personal being - just to fuse with a woman and fertilize her eggs. There's nothing left of him. He's now one with a horror of a female whose giant slicing jaws could terrify the Joker.

I'm certain there are men who feel like they are married to the deep sea anglertrosity. We've all seen it! We've all seen the same thing in some human relationships, and I believe the anglerfish is supposed to serve as a warning to men and women alike. "Men, don't date that chick. Don't get involved with her, or you'll look exactly like the male anglerfish. It's not pretty." Women too. "Girls, is this what you want your marriage to become in 10 years? Is dominating your husband a good thing? Is this what love looks like?"

There are relationships in the animal kingdom that serve as a picture of what really bites, and there are others that portray what we *should* be. The husbands of the female anglerfish lose their minds, bodies and souls to a horrific beast. Female lions do all the work while male lions lie around and soak up the sun. Male lions and male grizzlies will kill their own children. Not cool, guys. Wolves, though. Wolves are family animals. They take care of each other and are fiercely loyal. French angel fish stick close to each other and fight to protect each other. Penguin fathers are devoted and nurturing.

Even cockroaches can be more faithful than some humans. They're gross and we hate them, but cockroaches go home to the same nest every morning. "Hey baby, I'm sorry I only brought home a few crumbs. They went and bleached the kitchen floor again."

So, really. The point is this: we humans think we're so great. And we are. But, we would be better off if we treated each other the way stinkbugs treat their kin. And if we cheat, we're worse than a lousy cockroach.

Go to the ant, you sluggard; consider her ways, and be wise:
Proverbs 6:6

Chapter 28

Which God?

Jared: It's not ground beef. It's 'taco filling.'
Dr. S: And that's why it tastes so good. It adds a level of mystery to your food, and we all want more mystery in our lives.

I watched the white lines flash down the center of the highway as we zoomed along through the dark night. We had finished the last Odyssey of the Mind competition of my high school career in Yakima, Washington, and Mr. Carroll piloted the van filled with overly-creative youths back home to Spokane.

I sat just behind Mr. Carroll's right shoulder. It was dark in the van, and most of the kids in the back had dozed off. I had traveled across the state many many times on holidays to go visit my father, and I knew the terrain outside our windows despite the darkness. The central Washington badlands stretched to the horizon north and south of us, but the only source of visible light in all that quiet blackness came from our own van. As I watched the white lines fly past - dash, dash, dash - I listened to Dan Meyers describe the mystical powers of Buddhist monks from the front passenger seat.

"I watched a show where they blindfolded this guy and gave him arrows and a bow," Dan marveled. "He was able to turn and shoot the bullseye on targets. One after another without missing one."

I didn't doubt the Buddhist had been able to perform this feat. I didn't assume it had been staged; I simply questioned what specific powers the Buddhist had tapped into. Long hours of practice and muscle memory? Inner sight? Had the Buddhist really developed a connection with the Universe through meditation or were there other forces at work in his art?

The conversation eventually wandered around to the matter of God and the nature of reality, and I gave Mr. Carroll my personal views on the matter from my position behind his right ear.

"Yes," he said gently after I'd stated my beliefs. "But how do you know your God is the actual God? There are a lot of different religions out there with different views. How do you know that yours is the one that is true?"

It was a good question. How could I know?

I didn't tell Mr. Carroll, but that puzzle stuck with me for a long, long time. I graduated from high school and entered into my last summer before college. Month after month passed, and the question continued to haunt me, because I didn't know how to answer it. It created a real faith crisis for me. I couldn't even pray, because I didn't know which deity was listening. It was only in pure honesty, when I would say, "Lord, I don't know who You are…" that I connected to the God of my childhood.

I turned 18 that summer and faced the same general struggles that are common to young adults. There came an intense moment of frustration when I stood upstairs in my sisters' bedroom, staring out the window at the barn and fields beyond. "I'm not going to do this anymore!" I insisted. "I'm just going to go do what I want to do!"

I said it with all the intensity of youthful desire and energy, the longing to be free and not required to follow anybody's rules but my own. I'd decide what was right for me. I'd do what I felt like.

As soon as I spoke the words, I saw the emptiness and uselessness of the self-gorging I'd imagined. That fiery emotion extinguished in the next second as I came face to face with the hollowness, the emptiness of satisfying my flesh to the death of my soul. In the end, it wouldn't be worth it. It just wouldn't. I didn't know *who* God was at that moment, but I knew He *was*. In the deepest part of my heart I knew it, and I was stuck.

It's no good to worship false gods, no matter what they are. There are plenty of spiritual forces out there pretending to be gods, but the only God worth worshiping is the Most High God, the God that Is. The God of Truth. Who is that God, though? Which God is He?

One of my favorite commentators on idols is the prophet Isaiah. He describes a man who cuts down a tree and uses half of it for food and warmth and then uses the other half to make an idol. Isaiah makes a big deal about it, because it's obviously messed up. The prophet can't get over the nonsense of it, saying:

> *He burns half of it in the fire; With this half he eats meat; He roasts a roast, and is satisfied. He even warms himself and says, "Ah! I am warm, I have seen the fire." And the rest of it he makes into a god, His carved image. He falls down before it and worships it, Prays to it and says, "Deliver me, for you are my god!" ...And no one considers in his heart, Nor is there knowledge nor understanding to say, "I have burned half of it in the fire, Yes, I have also baked bread on its coals; I have roasted meat and eaten it; And shall I make the rest of it an abomination? Shall I fall down before a block of wood?"*
>
> Isaiah 44:16-17,19

What false gods have we invented without realizing it? In our society we don't usually worship statues, but we still create our own gods. We invent the gods we want to serve, making them up in our own heads, forming them to look the way we want them to look, whatever makes us comfortable. Utter silliness. Any god of our own invention is ultimately as worthless as an idol carved from wood. Mr. Carroll's question had disrupted me, but only because I wanted to serve the God who *actually* existed - whoever He was.

Months passed and I struggled on in my aggravated existence, unable to turn my back on God, but left with the dissatisfaction of hanging in empty space with no solid spot to place my feet. I felt like I hovered in midair with a giant chasm below. I didn't fall, but I hated it just the same; I wanted to see good, solid earth beneath my shoes.

That fall, I started college a few miles from my father's house in the Seattle area. Baron and Shadow had gone to live with Dad,

Which God?

which was great, because I could visit them on weekends. Walking through the boys' room on a Saturday, I saw one of Shadow's school notebooks open to a drawing. He had sketched a man's profile, and I stood by his desk, astonished at my brother's detail. He didn't just pencil out the man's cheeks and brow ridges; he showed every stretched and tortured muscle and sinew. I could almost feel the graphite blood pumping through the flesh of this intense human being that had come to life on the inside cover of a 14-year-old kid's notebook. Shadow had invented facial muscles that didn't exist, and he needed to move the man's eye backward, but it didn't matter. The boy had skills.

Figure 25: An 8th grade sketch by Shadow Truman.

Baron filled his workbooks with drawings too, but while Shadow's were full of intensity and anguish, Baron's sketches showed cartoon characters or peaceful woodland scenes. He drew rustic barns and people leaning against trees, reading. He drew children staring up through sequoias.

Both of my brothers had developed their talents. Both failed to do their school work. They were twins with the same genetic predisposition for violence, the same intensity and hatred for injustice. They'd spent their childhoods fighting bullies, and neither ever lost a fight - except to the other. Yet the differences in their two personalities popped out in their art. After looking at his drawings, one could believe that Shadow grabbed fire pokers and chased his sisters around the house in a rage. It made sense that he'd go on to wrap up terrorists in Iraq, that he'd be able to walk into a room and shoot the guy holding a hostage without fearing he'd miss and hit the hostage. From his drawings, one could believe that Baron

Figure 26: A pencil drawing on a note card, made by Baron Truman..

would become a generous evangelist who just wanted some peace in the painful world he faced. I saw a distinct difference in their two personalities even then, back when they were 14.

"Wow," I thought. "You can tell a lot about Baron and Shadow by their art. You can tell a lot about people by the things they make."

And it exploded in my mind.

You can tell a lot about *God* by what He's made too.

!!!

The Apostle Paul had thought of it long before I did, saying:

> *For since the creation of the world His invisible attributes are clearly seen, being understood by the things that are made, even His eternal power and Godhead, so that they are without excuse...*
>
> Romans 1:20

With that thought, I sat at a table and scribbled out the different things the world told me about its Creator. If I was going to answer Mr. Carroll's question, this was the best way I knew to start. Here's what I came up with.

The earth spins steadily. The sun rises and sets every day.

We can depend on the laws of physics. We can predict things, because the universe behaves in a consistent fashion that allows us to write laws and mathematical equations that continue to work year after year. The atoms in my hands stick together. They do not fly apart every once in awhile for the sick fun of it. The individual components all fit together in a way that makes sense. They do what they're supposed to do in a multitude of biochemical reactions that take place over and over every day, month after month, year after year. We can't change those laws to suit our whims - we'll fall off a cliff if we jump and flap our arms. However, we can use those laws to make airplanes fly and to soar on the wings of squirrel suits.

1) God likes order and fidelity.

We have to eat food to survive, and a wide variety of foods taste good. Sweet and salty and savory, food is good. And it's abundant. We can help by preparing the soil and planting seeds, but food grows by itself all the time. Millions of pounds of fruit fall to the ground every year unharvested. Little flowers sprout on mountaintops where few people will ever venture. We find cocoa beans and coffee beans and sugar cane and milk on the planet Earth. It's like God knew we needed mocha shakes. What a delightful place.

2) God is vastly generous and makes good things.

It can take 3-5 months to walk across America. Planet Earth is vast compared to our bodies. Yet, our planet is only a speck of dust in the universe. Stars are placed so far out in space that we're still only now seeing the light from them, and God is bigger than the universe.

3) God is huge beyond comprehension.

We live in a corrupted, broken, dying world. Yet, even though our world is doomed for destruction, the cells and organs in our bodies all have a purpose. I can live without my appendix and my adenoids, but these organs each have a reason for being there. They do, and it's better to try and keep them in our bodies if we can.[1] Even plankton and *E. coli* exist for crucial reasons. What's more, the body can heal itself when it's injured. It can fight off invaders. It can adjust to cold and heat, feast and famine.

Have you noticed that even the worst things can serve vital functions? Earthquakes bother us by knocking down our brick buildings, but tectonic activity is important for building mountains so that the earth has water stored as snow that melts in the summer, feeding our rivers and lakes. Hurricanes bother us by breaking our windows, but they are important for spreading seeds and other life to various parts of the world. And when we have to deal with earthquakes and hurricanes, we stop being self-centered, self-absorbed boobs who whine about our hub caps. We reach out to each other and serve one another, and when we do that, the world becomes a lovelier place.

4) God is purposeful and wise and He plans ahead for varying inevitabilities. He can even use destruction for good.

There is pain in this world. Sometimes people live in horrible pain for extended periods of time. However, this happens when *normal* processes break down. Pain is the result of something gone wrong. Severe pain is caused by disease or genetic disorders or injury. It doesn't come standard. The standard model is being pain free. When pain occurs, it's an important form of communication that informs the brain when something is wrong. Biological creatures need pain or we'll destroy ourselves without even noticing.

Even then, there are plenty of plants available that reduce suffering and stimulate healing. When we're seriously damaged, our bodies go into shock to protect us from excessive anguish. We have adrenal glands to give us an extra boost when we have to run or defend ourselves or pull cars off of little kids. When mice are attacked and dying, endorphins flood their tiny bodies to replace

---

1 Cf. Sanders, N.L., Bollinger, R.R., Lee, R., Thomas, S. & Parker, W. (2013). Clinical observations and predicted function of the appendix explain the relationship between appendectomy and C. difficile colitis. *World Journal of Gastroenterology*, 19:5607-5614. Also Cf. Boonacker C et al. (2013). Immediate Adenoidectomy vs Initial Watchful Waiting Strategy in Children With Recurrent Upper Respiratory Tract Infections: An Economic Evaluation. *JAMA Otolaryngol Head Neck Surg*. 139(2):129-133.

their terror with peace.

5) God is merciful and is willing to ease our suffering.

Our brains have the capacity to understand jokes and to laugh, to tease and have fun. We have the capacity for love and joy. We take these things for granted. They didn't have to exist. We can feel anger and frustration and injustice. We think and communicate and play. Even birds and dogs and horses and fish can play and have fun. We create. Our brains allow us to imagine and invent and compose music and poetry and build incredible ships that cross oceans and sketch pictures of little children under the redwoods.

6) God is creative and has a personality. He knows anger and injustice, love and joy and laughter.

We need to reproduce to survive as species, and our bodies are put together so that sex is something we desire. It's kind of a weird process. It's an absolutely weird process. I've never gotten over how weird it is. But it's fun and causes bonding, and when used wisely it communicates love and affection. Sex is not the only form of physical love, though. Hugs and kisses and a warm hand on our shoulder, these are things we can enjoy and share in the daily course of our lives. In fact, we humans *need* physical touch.

7) God made us to be close, to bond with each other, to give and enjoy affection.

We humans do horrible things to each other. We behave selfishly. We are petty and stupid and embarrassing. But. We also have a built-in conscience. We know selfish, destructive behavior is wrong. We know that love is good and cruelty is bad, and that's why we like to watch movies where the good guy wins and the bad guy loses. We make poor decisions, but not because we don't know better. We naturally understand concepts of right and wrong – that spot in our brain exists. We have the freedom to choose what we're going to do, but we also recognize that our actions have consequences. Better yet, we understand both justice and forgiveness, and the importance of both.

8) God understands right and wrong and consequences, justice and forgiveness. He understands these things better than anybody.

These are not the qualities of silly Greek gods as emotional and petulant as we are. They are not the works of a universal spirit with no personality of its own, or a hateful god, or an irresponsible supernatural child crashing cars together, or fairies or nymphs or any other small, personal spirit.

When we spend time with kind, wise people who love us, who are quick to laugh and quick to forgive, it's awesome. It's wonderful. It's the best times of life. If we simply chose good and rejected evil all the time, this world would be a delightful place.

The God who did all this is big and careful, wise and unimaginably patient. He is faithful. He likes order and beauty and diversity. God is a master artist. He paints a moving, living picture using exploding masses of gas far out in space and microscopic cell-life characters. He's fun and confusing and amazing. He didn't make us to be slaves. He made us to question and investigate, to create and invent. He made us to live.

That helped me a ton, Mr. Carroll. That's who God is.

There is a story - I haven't found its source - about a man named Chaluba who came from a family of idol makers. They worked with wood or clay or various metals to create the gods that people bought to worship. One day Chaluba said to himself, "I make these gods with my hands. Who... who is the God who made my hands?" He began to seek that God, and from what I understand, he found Him and went on to tell his story to the world.

Who is the true God? It's one of those all-important kinds of questions to get right. He's the God who made my hands. And whenever I wonder whether God's even paying attention, I look at my hands and think, "Yep, the muscles and tendons and blood vessels still have their purpose. The atoms haven't flown apart yet. God is wise and faithful."

Chapter 29
Life Presses On

Dr. Stillwell took us to pizza after we climbed out of the Grand Canyon. Wonderful, hot, gooey pizza after a day of hiking.

"Have you ever had Canadian bacon and pineapple pizza before, Flash?"

Dr. Stillwell stiffened and glared at me across the table.

I cringed. "I mean, Dr. Gurden?"

"I'm sorry," I whispered to the geologist. "He *told* me to call him, 'Flash.'"

"What's Canadian bacon?" asked another student.

Dr. Stillwell didn't care that Dr. Gurden had told me to call him "Flash." I was doing a poor job in my leadership of the other campers. Dr. S. got miffed at me a lot for being a bad example. I threw snowballs. I lagged behind to take pictures of Jared. I giggled when Emily mocked Dr. Stillwell, and I called Flash by his nickname.

At the Grand Canyon, Flash was a bad example too. He took film footage of my snowball throwing, and he lingered at a nice view and fell behind. But, Dr. Stillwell didn't go up to Flash and grab him by the backpack and drag him along like he did to me.

Dr. Gurden is one of my heroes. He's tagged along as Dr. Stillwell's co-teacher on every geology trip I've taken, and he's the best. He's calm when driving behind Dr. Stillwell's van without a clue where we're going, even when it involves driving into the desert with no destination or making U-turns in the middle of Boston. Dr. Stillwell can handle all the stress; Flash goes on the trips to enjoy himself.

As I write this in June of 2015, I'm sitting in Flash's living room. He and his wife Amanda have just taken off to the hospital because her water broke. Before they left, Flash and I waited in the kitchen while Amanda used the restroom one last time.

"What if... what if she gives birth in the bathroom over there?" I asked Flash. "Are you prepared?"

He shrugged, unworried about it. "We'd save some money?"

We both laughed. We were giggly anyway, because the baby is coming, and it's exciting.

I told him, "All you do is, you boil some scissors. You get some shoelaces, and you're good. Superglue is awesome."[1]

"You know, it's strange where life takes you," Flash said thoughtfully. "If I'd have been told when I first met you, Amy Joy, that you would be here holding down the fort while we went to have our second child, I would have... I probably... I would have believed it."

We started laughing again.

He finished, "I'd have thought, 'Yeah, that makes sense.'"

Today is June 25, 2015, and the Gurdens are off to bring their son into the world, while I stay back with their three-year-old little girl. Flash called me at the Caerphilly farm about 9:00 p.m., and I drove over, tears welling in my eyes. New babies always make me cry. I felt the immense privilege of sharing a treasured moment in my friends' lives.

It was a beautiful evening to boot. The afternoon rains had drifted off, and the last summer rays caught on the mist rising from the road before me. As I drove through acres of green fields, lightning bugs danced like beacons of joy across the tops of the young corn plants. It was a gorgeous day today.

It will probably be morning when Flash returns, but when he does, I get to run down to the hospital to hold the little guy myself. I admit, I wish Amanda had let me deliver the baby here in the living room.

---

1 There's more to it than shoelaces and superglue. Don't sue me.

Chapter 30

The Life Inside Me

My knowing-in-my-knower continued with Dr. Stillwell for the next several years.

"You haven't visited me!" Dr. Stillwell accused me one day in early 2012.

"Well, you've been so busy!" I said. "Yesterday morning you were inundated with people. And then about 10:30 you thought, 'I'd like to see Amy Joy,' and you missed me for about 30 seconds. I was going to stop by, but then you were inundated again, and it was too late."

I didn't observe any of these things with my physical eyes; I was blocks away when I felt them. Still, he nodded, because it was so.

Rejection has been my baseline since early childhood, drilled into me by repetition. It's what I expect. I'd have never guessed that my tired, overworked geology professor thought of me in the middle of the morning when he was focused on so many other things. He doesn't multi-task, and he gets mega grouchy when he's busy. I offered that description to him only because I had absolute, full confidence it was true, even though I had no physical way of determining it. I simply knew it, and that gave me the courage to say so.

Later, I reminded him of that particular day, and he grumbled, "Well, that happens all the time."

Why would I suspect that if you say nothing, Doc? I never know who is in charge. Today I might see my dear friend Dr. Henry Jekyll, but maybe the evil Ed Hyde who resents me will show up.

Since moving away from school, this peculiar connection has not gone away. Since 2012, I've said to myself, "Hmm… I'd like to give Dr. Stillwell a call." But it's no good. I can't call him. "Oh…

nope. Nope. It's not a good time." I can sense it. On two occasions I called him anyway, thinking I would leave him a voice mail message. Both times he answered, irritated, and said, "I can't talk right now."

On the other hand, I'd say, "Okay. He's busy. I won't bother him for awhile." Halfway through the next week, I'd get the urge to give him a ring. So, I'd phone him up and he'd answer, "Amy Joy! You called at just the right time! I just walked in the door."

Over and over.

It's been so consistent, I've wanted to figure out how we could frame an experiment. Take down dates and compile data. Of course, as soon as I decided to start, months went by with no good time to call. Months. He was too busy and focused on deeper problems.

In February of 2014, I visited the Caerphilly home to help with the 45 baby goats ready to be born. We had 14 kids spring out in one day, and Flash and his wife brought their little girl to cuddle with the soft goatlings and watch them jump around. Dr. Sytil stopped by with her children, and Dr. Manchester's wife brought their youngest daughter. The Caerphilly goat farm got popular at kidding time.

Of course, I popped by Dr. Stillwell's office while I was in town. I hadn't called him in a long time and confessed to him, "I think I've lost my superpower. I can't tell anymore when it's a good time to call you." Then I countered, "Or maybe there's just been no good time to call."

He nodded. "Maybe there's just been no good time to call." It was his last semester as department chair, and I don't think he had any breathing room.

It's annoying, because I really would like to set up an experiment. Make predictions. Have Dr. S. make observations as well. Have him write down moments when I could have called but didn't. I don't know what's going on in his world. I only know that when I've called him because I've had the urge to do so - "Okay, *now*!" - he's invariably answered and said, "Your timing is perfect!" That's all I know.

In September of 2014 we had a horrible fight. It was huge, the biggest fight we'd ever had. Even then, even *then* he didn't deny that I consistently had perfect timing.

"You know my schedule!" he roared.

In that heated, furious moment, he didn't deny it, and that pleased me. It told me that he hadn't been playing me. He hadn't been saying, "You called at just the right time!" to be polite or nice or friendly. He knew the truth about the events of his life thousands of miles from me, and I didn't. I only knew how he reacted to my phone calls.

It's the weirdest thing. I don't know what to make of it except that it's been accurate. And his schedule has nothing to do with it.

After we returned from Arizona and Utah in 2011, I started my research on bryozoans. Dr. S. still didn't have much time, but he stopped in now and then to check on me. We had a particularly good, long talk one day that summer. Dr. Stillwell said some things that had been on his mind, and I said some things that had been on my mind.

"What is it about me?" I asked him again. "Why is it that you enjoy my company so much?" I'd asked this question of him a couple of times. He always gave some lighthearted answer that didn't get at the guts of it. Had he lost his children? Had he lost a close friend? What hole did he have?

Dr. Stillwell is the one who hides himself. He's pulled a protective denim over the innermost parts of his heart, and he doesn't share. On this occasion, he got closer to being serious with me than he'd ever been before. In that quiet moment, he thought about it and said, "I don't know. It's just… there's this LIFE inside of you."

Ah. I nodded. That was it after all. "That's what I thought. That's what I thought it was."

He looked excited for a moment. "Is that what you get from me?"

I shook my head. "No." I didn't bother him with it right then. I didn't want to tell him that he was a hollow shell. I didn't want to say those things to him.

We're dear friends, even if he does feel empty inside. I'm just looking forward to the day when he has his own joy, when he himself is full of LIFE. I'm looking forward to the day when I feel a deep comfort from him, a warmth and contentment that comes from his innermost being. I look forward to that day.

In the meanwhile, I love that old codger. And he loves me. And I think that's pretty cool all by itself.

During the summer of 2011, I had a dream about Dr. Stillwell, and in it he was singing a Chris Rice song. It's a sweet song, and I think it's what many of us would sing if we stood in the quiet dark of night with nobody but God Himself. Despite his grouchy exterior, I sometimes wonder if it isn't the song of Dr. Stillwell's heart.

None of us knows, and this makes it a mystery,
If life is a comedy, then why all the tragedy?
Three-and-a-half pounds of brain try to figure out
What this world is all about,
And is there an eternity? Is there an eternity?

God if You're there, I wish You'd show me,
And God if You care then I need You to know me.
I hope You don't mind me askin' the questions,
But I figure You're big enough.
I figure You're big enough.

When I imagine the size of the universe,
And I wonder what's out past the edges,
Then I discover inside me a space as big
And believe that I'm meant to be
Filled up with more than just questions.

So, God if You're there I wish You'd show me,
And God if You care then I need You to know me.
I hope You don't mind me askin' the questions,
But I figure You're big enough.
I figure You're big enough.

To Be Continued…

APPENDIX

More fun. So much fun!

1) Angels and Witnesses.................. pg. 197
2) Birds in Stone .... pg. 200
    • Calibration .... pg. 202
    • ...Versus Code .... pg. 204
    • Digital Programming........ pg. 207
    • Blind Watchmaking.......... pg. 210
    • Twisting Up Proteins........ pg. 212
3) A Simple Light-Sensitive Spot .... pg. 221
4) Entropy...................................... pg. 231
5) Of Mice and Men .... pg. 233
    • Of Women and Men .... pg. 237
    • Of Chimps and Men .... pg. 238
    • Of Pigs and Men .... pg. 241
    • Convergent Evolution ..... pg. 242
    • Of Fruit Flies and Men .... pg. 243
    • Junk DNA   pg 244
    • Ophan Genes: .... pg. 246
    • Some Predictions .... pg. 248
6) What Are Bryozoans, Anyway? ... pg. 251

1) Angels and Witnesses

Notice how fiery messengers appear and disappear at will in various Bible accounts? Angelic beings are described in passages written by different people over the centuries, both in the Old and New Testaments, and we can find both holy and demonic entities all over the different world's religions.

These beings are physical – they have physical qualities – and they appear like men when they enter our dimension. In Genesis 19, two angels arrive to escort Lot and his family out of Sodom before it's fire and brimstoned, and those two angels are capable of both eating and taking people by the hand. They're obviously not normal men, because they're able to instantly blind the violent crowd outside, and they explain that they've been ordered with the city's destruction.

In Daniel 10, a warrior angel appears to Daniel after three weeks of fighting past enemy forces. He looks like a man, but he glows with a fiery brilliance, and his mere presence causes Daniel to faint. We think of angels as spirits, but they have physical bodies just like we do when they enter our plane of existence.

Of course, the Bible doesn't tell us, "Hey! There are eleven dimensions!" It doesn't say that. It doesn't try to explain the greater nature of the universe or give us mathematical equations to neatly provide a "Theory of Everything." The Bible simply takes those extra dimensions for granted.

I didn't think about it when I was young, but Jesus also demonstrated greater dimensionality after His resurrection. Before He died, Jesus had to get into boats and walk and ride donkeys to get from here to there. After His resurrection, He disappeared and reappeared at will, just like the angels. Hebrews 1 makes clear that He's not an angel, so don't be confused. It's just noteworthy that His rising from the dead came with greater dimensionality.

Near the end of Luke 24, Jesus appears to the frightened disciples in a room where they're hiding in fear of the authorities. He doesn't walk through the locked door; He simply appears in the middle of them.

Luke 24:39 tells us Jesus said, "Behold My hands and My feet, that it is I Myself. Handle Me and see, for a spirit does not have flesh and bones as you see I have."

Jesus is physical and tangible here. He eats some fish and honeycomb in front of his disciples. (In fact, He's always eating after the resurrection, which is comforting. It's clear that we get food in eternity.) Even though the post-resurrection Jesus is a physical being, He's able to enter and exit locked rooms without a problem.

Jesus gets especially personal during one of these appearances. Thomas wasn't present when Jesus showed up on Resurrection Sunday, and this doubting disciple didn't believe his companions when they told him about it. Thomas was a good old skeptic. When Jesus appears again, He allows Thomas to feel His wounds, to demonstrate that He's the same Jesus that Thomas saw crucified and is very much alive.

> *And after eight days His disciples were again inside, and Thomas with them. Jesus came, the doors being shut, and stood in the midst, and said, "Peace to you!" Then He said to Thomas, "Reach your finger here, and look at My hands; and reach your hand here, and put it into My side. Do not be unbelieving, but believing."*
> -John 20:26-27

Thomas has followed Jesus for three years, and the two of them are friends. Jesus doesn't tell Thomas, "Don't look at the man behind the curtain!" No, he urges Thomas to feel the wounds himself, to handle Him and see for himself the truth of the things the other disciples had told him.

The Bible is about faith, but Jesus didn't insist that Thomas had to believe without evidence. Evidence is important. Multiple witnesses are important. The books of Deuteronomy and Matthew tell us that events should be established by at least two or three witnesses. The Bible follows this rule itself, giving us two sets of accounts about Israel's kings (in Kings and Chronicles), two sets of the Law (in Leviticus and Deuteronomy), and four sets of Gospel accounts about Jesus (in Matthew, Mark, Luke, and John). God's primary principles and promises show up repeatedly across the Bible in books written by men who lived centuries apart.

When I relate stories about miracles, or exorcisms, or other supernatural events, I try to get at least two accounts from different people who were there. The more witnesses, the better. Even if I thoroughly believe a person, I always work to find other people who were present to get additional perspectives. The fact that there are four Gospel accounts is important, because they all tell a similar story, but we get a view of the life of Jesus from different perspectives.

The fact that Jesus walked through walls is established by both Luke and John. Luke was a historian who made an investigation of the events surrounding Jesus Christ, and John was an eyewitness. Luke went around interviewing people, and John said, "I was there." They both describe Christ's odd habit of showing up without entering through doors. When Luke talked to the men who walked with Jesus on the road to Emmaus, they said that Jesus disappeared right in front of them at the dinner table.

The universe is a strange place. Multitudes of people have experienced encounters with both good and malicious spirits. It's easy to dismiss the stories of other people, but it's more difficult to ignore the invisible entity that throws dishes across your own kitchen.

The Bible doesn't blatantly tell us that there are more dimensions, but we get clues from the consistent stories its writers give us. And while these stories might sound like the stuff of fantasy and fairy tales when read by skeptics, they are consistent with the nature of the universe according to theoretical physics.

2) Birds in Stone

The day we walked the Grand Canyon rim, Dr. Stillwell and I examined flat bird sculptures fashioned into the pavement at a visitor's center in northern Arizona. We wandered back and forth across the pavement, studying the different species at our feet. The sun shone down on us in that courtyard, baking the members of the stone flock around us. Soon, Dr. Stillwell and I stopped above a "CALIFORNIA CONDOR" sculpture sunk into the pavement. Its wings stretched out past me on both sides.

"It's fake," Dr. Stillwell gazed down at the flat stone bird.

"No," I said. "No, I think it was etched by the rain. It's really pretty simple. It's just the result of natural processes. The letters… there are only 15 of them…."

"And a bunch of monkeys on a typewriter," he started to recite.

I finished it for him. "Would write Shakespeare. Actually, in real life they just punch the letter X a hundred times and fling poo at each other."

"Hmm," said Dr. S. "That sounds like *The Washington Post*."

I laughed out loud.

The professor and I both recognized the clear and obvious fact that human beings had made those pavement birds. There were no doubts shooting down our neural pathways. Human artists had formed those creations, and rain had little to do with it. If anything, rain promised only to erode and damage the art over time.

We *Homo sapiens* are pretty good at discerning what is natural from what is human-made. For the most part. We can gaze up and see pictures in the clouds or find the shapes of animals in the cracks of a rock outcropping. Our brains create meaning from random shapes, but we discern they're just clouds and rocks after all. On the other hand, Dr. Stillwell and I both recognized it would be silly to suggest that rain – an eroding, etching natural process – could carve the picture of the California Condor and the letters labeling it. Rain has the ability to etch, but it's an insufficient cause for writing letters under the picture of a specific bird.

Yet, flat stone birds are relatively simple! They have no moving parts. They don't perform any active functions. They don't grow and fight off infection and heal themselves and reproduce other miniature rock carving duplicates of themselves. Something as modest as a flat bird outlined in the ground, something that uncomplicated; we knew it didn't form by rain carving out the rock. We recognized that there was a mind and planning and creativity behind every one of those birds and the letters telling us their names, and we didn't have to watch the artists working to know it.

William Paley famously made "The Watch and the Watchmaker" argument for God's existence in his 1802 book *Natural Theology*. He stated that if we kick a rock, we can't say much about how it got there (geologists would disagree), but if we kick a *watch*, we can know it was made by a watchmaker, because it demonstrates purpose and design. It doesn't even matter if some of the watch parts are superfluous or if the design is good. It's obvious from its intricate moving parts that an intelligent agent had constructed the watch before it was dropped into the dirt to be kicked. It's a 200-year-old "teleological" argument for God's existence from purpose or design found in nature. Paley concludes:

> Every indication of contrivance, every manifestation of design, which existed in the watch, exists in the works of nature; with the difference, on the side of nature, of being greater and more, and that in a degree which exceeds all computation. I mean that the contrivances of nature surpass the contrivances of art, in the complexity, subtilty, and curiosity of the mechanism…[1]

Paley was right about the engineering we find in nature; it's better than anything we come up with. In fact, human inventors have long studied natural structures to figure out how to design things better. Architects wanted the Eastgate Centre in Harare, Zimbabwe to

---

1   Paley, William, (1802). *Natural Theology; Or, Evidences of the Existing Attributes of the Deity, Collected from the Appearances of Nature*, Chapter III.

remain naturally cool in the African heat, so they designed it using research into the structures of termite mounds. Velcro was inspired by the burs that stick to our socks. Richard T. Whitcomb designed airplane wingtips to curl upward in the 1970s after noticing that the wingtip feathers of birds bent upward. He discovered that winglets significantly reduced the effects of drag.

One of my favorites is the long, bug-grabbing woodpecker tongue, which is stored behind and over its skull in a hyoid apparatus. The barbs on the tongue are great at pulling prey out of little holes in trees, but the tongue-storage serves as an important part of the bird's shock absorption system. Woodpeckers repeatedly slam their beaks against tree trunks, each impact hitting at better than 1000 times the force of gravity. Researchers have studied the woodpecker skull structure for ideas, and on February 4, 2011, a *NewScientist* headline declared "Woodpecker's head inspires shock absorbers." There is clever engineering in nature, and everybody recognizes it.

Paley's point was that engineering requires a cosmic Engineer, and thus the brilliance in nature - which is much greater than our own engineering wisdom - points us to the reality of God's existence. Teleological arguments are old, much older than Paley, and Psalm 19 and Romans 1:20 both tell us to listen to them.

Paley's "watchmaker" argument is brushed aside these days, however. It's even laughed at. Richard Dawkins used his 1986 book *The Blind Watchmaker* to answer Paley and explain the view that natural selection has blindly, but steadily, created the engineering we see in nature. Natural Selection is the new lord of creation. Throughout the biological sciences, we find it given credit for everything. Its vast power is credited with every work of genius, every sophisticated collection of biological machinery in the world. God is not necessary.

CALIBRATION...

I appreciated Dr. Bob Manchester's take on the situation. Dr. Bob sat across from me at lunch during the spring of 2011 and told me he believed that God set it all up at the beginning. He believed

God defined the laws of physics and tuned-up the universe so that life could exist and develop.

Dr. Manchester is bright. He's a dedicated biochemist and a good guy. I have a tremendous amount of respect for Dr. Bob and trust his judgment. However, if he's right, then it should become obvious that evolution happens as a matter of course. It should be obvious that it's the normal case, that new functions can and have evolved from scratch over and over and over and over during Earth's history.

That's the huge question. Is it even possible for biological machinery to evolve from scratch? Is that how it was set up after all, or are the world's biologists making incredible assumptions they have no way to test? In reality, was it biologically necessary for the major animal forms to have been created fully functioning at the beginning? Which is it? How can we determine the best answer to that question?

The universe has been fine-tuned and set up for life so that the laws of physics and chemistry work the way they do. That's obvious to everybody, and I recommend John Barrow and Frank Tipler's 1986 book *The Anthropic Principle* for minute details on this matter. It's an older book, but the nature of the universe hasn't changed in the past few decades.

Despite the overwhelming observational support for the Anthropic Principle, John Barrow still insists that God isn't necessary for the fine-tuning of the universe. He suggests that there are a multitude of possible universes, and this one happened to work for life. Everybody loves the multiverse idea these days (sigh). They believe that if there were infinite parallel universes out there, our particular world could show up by statistical probability. That is, if there were enough universes, then every optional movement for every atom would take place, including the interactions that created our version.

It's very clear the universe in general and Earth in particular are just right for life. That's undeniable. My huge question is this: Earth is just right for life, but is that sufficient to *build* life from scratch? An aquarium might be set up to sustain the life of clown fish, with the right temperature and oxygen content and pH and salinity and food availability to let Nemo flourish, but does that mean it can build clown fish from bits and pieces of material floating in the water?

That's the question!

Frankly, we don't know if there is a multiverse, and this is the only world we can observe. But, let's say the multiverse does exist. Even then, it doesn't matter, because *this* universe, this one we live in, is absolutely not set up to build life from scratch. In fact, this world breaks things down way faster than it builds things up, and I have physics behind me on that. But maybe I shouldn't be such a Debbie Downer, and I'll try to consider the possibility anyway.

For the purposes of biological machinery, I notice there are simple laws of attraction. Electrons flow like a river from negative to positive. Polar water molecules are attracted to other polar molecules, and nonpolar oils are attracted to other nonpolar molecules. Water and oil naturally repel each other; they don't mix. Chemical reactions happen naturally, over and over, in predictable, constant ways. These things are part of the way the universe is calibrated.

Because atoms stick to each other in a certain way, I have a solid laptop in front of me. I can type on this laptop because of the arrangement of its atoms, and because electrons move in a certain way, electricity flows through it, giving it life. All the constituent parts behave in a certain way repeatedly.

We can observe these things. We can see these laws of attraction and chemical reactions taking place over and over without any conductor visibly required to direct the symphony.

...VERSUS CODE:

However. Notice something? The laptop already exists. The programs run because they were already written. Like my laptop, my body runs and runs and runs, but it had to get here first, and it's far more complicated to create a body than to keep one running after it's already been set up.

"Because DNA is digital," I nodded at Dr. Bob across the table at lunch.. "It doesn't *mean* anything unless it already has a way to

Appendix

be read." I didn't know how to verbalize the significance of this to Dr. Bob. "It's like a watch. The numbers '6' and '3' and '0' in '6:30' only tell us something because we've agreed in advance that those symbols signify a time of day. It's not like a sundial, which is analog. A sundial works because of the angle of the sun in the sky."

"Yes," said Dr. Bob. "But even a sundial has to be set up and calibrated."

Wow.

I hadn't even thought of that, but he was right. I had long been impressed with DNA as a biological computer program, but even systems that aren't digital have to be calibrated. Good one, Dr. Bob!

There's something very important to note, something distinctly unique about this laptop before me. The words showing up on this screen are there because I typed them in. They aren't there merely because of natural laws of attraction. The letters on my screen line up in a specific order, complete with spaces and punctuation, because I'm adding them on purpose to communicates ideas. These letters you're reading are a code. They only have meaning because we've agreed in advance that the letter "s" says "sss" and "b" says "buh" and "m" says "mmm" and all the letters in various combinations form words that communicate the thoughts from my head. The letters on this screen only mean something because we were taught this specific code as small children. If I started writing in Klingon, very few of you would know what I said, because the Klingon symbols mean nothing to you. For instance:

⟨ꜰⱡ   ⱷꜰⱷꜰⱷⱷ'   ꜱⱡꜱ

I pieced this together, and I hope I did it right, because I don't know Klingon. I hunted down this Klingon phrase, and I've learned that there are three layers of translation that need to be given here.

First, we need to know that these letters transliterate to (I think) "Hab SoSlI' Quch" (which I don't know how to pronounce). Second, the words translated into English mean, "Your mother has a smooth

205

forehead." Even then, this still doesn't produce meaning unless we're familiar with Klingon idioms. We still have to get the third layer of the translation, which is where we learn that saying "your mother has a smooth forehead" is a big, fat Klingon insult. I've just disrespected you in Klingon.

Until we learn these symbols and their significance, a sentence in Klingon would be gibberish to us and communicate nothing. (Which is just as well. Rude!)

But, that's what makes symbols unique. It's why digital codes matter so much. There's thought and personality and purpose behind each of the letters I'm typing. Atoms are naturally bonding to form the shape of my laptop, with its keys, but these letters didn't form by natural processes due to automated atomic activity. They didn't even form by a cat walking across the keyboard or rocks falling on it. Those things would never create long strings of words that offer complex meaning.

There's even more to it. We don't only have to be taught a code to understand it. We require the very ability to decode in the first place, and we do! Our brain has the capacity for code-use. We can understand this specific message because we have learned to read and can understand English. We appreciate syntax and subtle differences in vocabulary.

Also, our *eyes* are able to collect the light and our *ears* are able to collect sound vibrations that convert words into electrical impulses from which our *brains* interpret meaning.

Our ability to comprehend code is remarkable because it requires several layers of translation. From light to neural impulses. From neural impulses to sight in our brains. From sight to translation of letters into words and words into meaning.

Letter shapes. Specific word spelling. Word meaning. Syntax. Eyeballs. Light. Neural connections. Brain. Life.

All these things are involved in your reading these words. Not to mention the multitude of interactions - digestion, oxygen transport, energy creation, immune armies, skeletal protection - all involved in keeping your brain chugging inside your body.

You should feel overwhelmed and unable to appropriately appreciate the marvel I just described. (I can't. I can't absorb it.) But that's the genius behind our language abilities, and it's also the genius behind DNA, and hopefully you already understand why I'm deeply skeptical that our universe built us from scratch on its own.

Let's keep going.

The DNA in our cells goes through several layers of code-wrestling too. DNA only means something because it was programmed in advance to mean something, and the body knows how to interpret that code. The letters are strung together in a specific order that directs the production of both proteins and the intricate timing of cell processes, and that code is read in a carefully choreographed manner.

That is huge. That is vital. That is the core of the whole problem with microbes-to-man evolution. We absolutely recognize that there are natural processes, natural attractions between molecules that cause certain chemical reactions to take place automatically over and over again. However, there is no known natural process that creates letters and codes without the intervention of a mind and thought and intelligence.

That's what I see in nature. I see a bunch of organisms programmed to function because of detailed, complex digital programs. Programs that are error-correcting. Programs that duplicate themselves. Programs that are flexible and kick into action based on varying environmental triggers. Programs that can even be swapped and exchanged. But, when they break, they don't work anymore. That's what I see.

DIGITAL PROGRAMMING:

I took a coding class at age 16, and that knowledge has come in handy at several jobs. The first thing we learned was that the binary code behind computer software is made up of two digits – 0 and 1. Every computer program on our desktops ultimately connects back to a series of 0s and 1s. DNA is a bit more complicated; it uses a digital code in the form of four nucleotide bases, Adenine (A), Thymine (T) (or Uracil (U) in RNA), Cytosine (C) and Guanine

(G), four letters - ATCG - instead of 1s and 0s. These four letters spell out the programs that are required for living systems to run. Those programs are long and detailed, and each cell in our bodies contains enough biochemical letters to fill a multitude of encyclopedia volumes. Stephen Meyer's 2009 book *Signature in the Cell* describes the digital nature of DNA in extensive detail, and I'm grateful for his careful, step-by-step trek through the subject.

Even at 16, I couldn't understand how DNA could be seen as evidence for microbes-to-man evolution. There's no biochemical reason for DNA to follow any specific sequence. The code is possible because the A C T G letters can join together in any order, as far as natural attractions are concerned. When I realized that biologists interpreted identical DNA sequences as evidence of common ancestry, I finally appreciated their point. Mice and humans and elephants and dragonflies all share certain sections of DNA code. Yes, they do. Except. Except, somebody still had to write all that code in the first place.

This is why I believe Intelligent Design theorists are absolutely legitimate and should not be dismissed as philosophically motivated. They are not using a "God of the gaps" argument, in which they say, "God did it, because we don't understand how something came to be." Instead, they conclude that engineers are the only sufficient explanation for highly engineered objects, and the only sufficient explanation for code is a code writer. Natural selection acting on mutations is insufficient to explain the encyclopedia-volumes of coding in DNA, exponentially, fantastically more insufficient than the erosional forces of rain water to produce "CALIFORNIA CONDOR" below the stone image of that particular bird.

Since 2012, I've been employed at an analytical chemistry lab where we analyze rocks for gold and silver, copper and lead and the full array of other elements. We have different machines ("instruments") that analyze samples for nearly every element found in Earth's crust. When I arrived, the woman in charge of making reports had been cutting and pasting the results from Excel spreadsheets produced by the instruments. Cutting and pasting! Not quite kindergarten all over again. She moved the lead and zinc and copper data into reports and emailed them off to customers.

This old system was tedious and error-prone, and it had to be updated, so our lab purchased a Lab Information Management System (LIMS) to get us into the 21st century. The point of the LIMS is to move data straight from the instrument computers to a database, from which reports can be printed out with the push of a few buttons.

No more cutting and pasting the results for 250 samples. It has become possible to spit results smoothly onto a report straight from the database, complete with the client's name and contact information. Great stuff.

Easy peasy, right? As simple as that sounds, the whole system has required a tremendous amount of programming. It did not come ready-made for our lab, and there were multitudes of little steps in the process toward making it work. The LIMS company sent a tech guy to plug in the specific instrument details so their results could be sucked into the database, but then I had to debug the whole system and write the codes for all our login worksheets and report templates. I won't bore you with the details, but I had a steady string of victories, "Yayy, it works!" over the course of 18 months.

Eighteen months.

For at least a year, the entire beast still didn't function. There were times my boss wanted to know what the [insert foul words] I'd been doing, because we still couldn't produce reports after I'd spent months steadily developing them. I tried to explain to him (and he is not an understanding man) that a multitude of tiny steps had to be fixed first, and the system wouldn't fully operate until those bugs were worked out.

We're now to the point where reports are generated straight from the database. Now I simply go in every day and fine-tune. Which is the easy part.

Our bodies are automated machines that run the same programs over and over. But survival of the fittest and natural selection are all about the fine-tuning. Which is the easy part. A species cannot fine-tune if it isn't living long enough to reproduce. Say that five times fast. DNA cannot be fine-tuned into existence. I cannot stress this enough: the programs can only be modified after they're already written.

The LIMS is finally a living system, and I'm in the stage of modifying report templates to fit all the many different analyses we perform in our lab. But, the process took a great deal of brain work and effort before the system functioned at all. From the outset, it looked like it would be a straightforward process. I had no clue how many details would have to be addressed.

The rain is a natural, eroding process, but it doesn't make letters in rock, writing CALIFORNIA CONDOR or any other words. Natural selection is a natural process that weeds out the weakest and encourages those best suited to an environment to survive. But, it doesn't write codes. It doesn't put the letters there. It just kills those creatures whose programs are not suited to the current conditions.

In my view, biologists fail to see biological systems with the eyes of computer programmers. They don't realize just how huge the problem is. They don't recognize the serious dilemma of having a system working in the first place before it can be adapted to an environment. There are real problems. Real problems. I still don't see the bridge across Grand Canyon that people keep insisting is there.

BLIND WATCHMAKING

I started reading *The Blind Watchmaker* in a local library when I was 19, and my first impression was that Richard Dawkins had talent at writing for the general public. I enjoyed reading his book, but it didn't take me long to think that he - as I said - didn't appreciate the magnitude of the problem. He argued that God was unnecessary for the brilliance of what only *looked* like cleverly engineered biological systems, that cumulative selection could form such marvels as the hemoglobin molecule or the human eye through a series of small steps. However, I didn't think he understood the nuts and bolts of what he was saying. It's not remotely that simple.

Let's take a little trip through this issue. Just a short one.

Part of an eye, Dawkins argued, is better than no eye at all. Eyes could evolve, he imagined, because simply detecting light was better than not detecting light, and just detecting shape was better than not detecting shape. Thus, every little additional advantage that an improved eye offered made it likely that it would increase chances of survival and allow those traits to be passed on to offspring.

That sounds good. It sounds reasonable.

Remember the 1986 version of Nintendo's *The Legend of Zelda*? At the beginning, somebody gives Link a wooden sword to fight off bad guys, which is much better than having no sword - just as the light-sensitive cells on a starfish arm are significantly better than no eye at all, because they allow the starfish to detect the location of the reef. Later, Link finds a white sword, which is like getting a lens to focus that light. It's much more powerful than the wooden sword. Eventually, Link finds the magic sword, which is like adding an iris to manage how much light enters the eye, or color-sensitive cones in the back of the eye. The wooden sword works to help Link survive until he obtains the white sword, which improves his survival chances until he finally earns the magical sword.

I followed Dawkins' logic, but I decided that he didn't appreciate the vast extent of the dilemma facing a blind starfish. It's not like little eye parts were just floating around in the water until one day, boom, light-sensitive cells got stuck at the end of a starfish arm so it could see the reef.

Light sensitive cells, like the ones in our retina, work because they contain specific photoreceptor proteins and those proteins aren't just floating around. They are created by a chain of amino acids that are placed into a specific order by DNA programming. Link found the new and better swords to fight Darknuts and Moblins because there was somebody who *wrote code* for each sword. Somebody wrote out the digital directions for Link and the Darknuts and Moblins too. They are all the inventions of a programmer.

Still. Couldn't every tiny, simple improvement work long enough to allow additional improvements, like Richard Dawkins has suggested? Each small change could allow a creature to survive

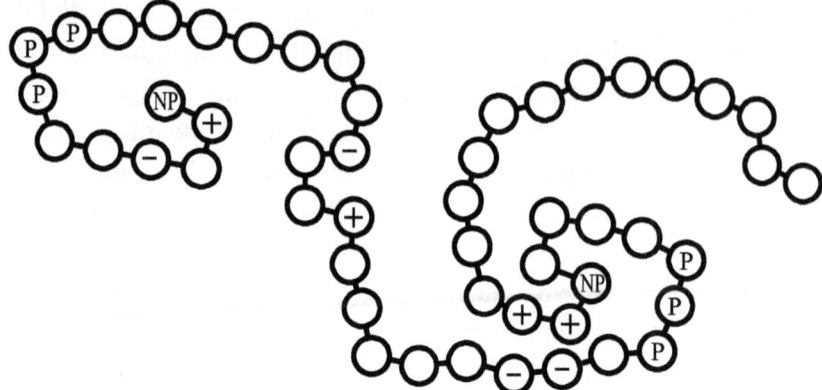

Figure 27: A strand of amino acids folding into a protein according to whether they are polar(P) or nonpolar(NP) (water loving / water hating) and positive(+) or negative(-).

better – giving it time to develop, right?

Okay. Let's consider that. Let's first look at what it takes to make a protein, and then let's examine the matter of the simple light-sensitive spot. If you didn't learn this in sophomore biology, I'm here for you.

Twisting Up Proteins

Light sensitivity is the result of certain proteins reacting to light. To understand this, we need to first know how proteins are made.

DNA directs a number of biological processes. One of its most basic purposes is to specify the *order* of amino acids in proteins, and the order of amino acids is important, because it determines protein *shape*, and protein shape is very very important, because protein shape is what makes proteins *do their jobs*. A protein's three-dimensional structure determines its function – just like the shapes of interlocking gears make a clock work correctly. The gears need to have precise dimensions, or they jam and the clock freezes up tight. Proteins aren't solid and hard like iron clock cogs, though. They are indescribably more wonderful than clockworks. They change shape in the presence of other pieces of biological machinery, as elegantly as Bumblebee transforming into a Camaro, so that they can do the jobs they need to do. They are living, moving gears and cogs. It's amazing.

Proteins are not just important for building strong muscles, either. There are a multitude of proteins working as enzymes and cellular machinery. They can be massive or they can be small, depending on the job they have to do. Some act as taxi cabs and others as chaperones. Some act as catalysts to speed up reactions, and some combine with dozens of other proteins to piece together into giant factories, like Legos. Living, moving factories.

Again, proteins do their jobs because they have a very specific shape, and the shape of the protein depends on how it folds, and how it folds depends on the individual amino acids that got strung together (see Figure 27). Each amino acid has characteristics – polar or nonpolar, a positive or negative charge, a certain size and shape – and when large numbers of amino acids connect together, like beads on a string, that string bends and twists up into its most natural conformation according to the way all the various amino acids attract or repel one another and the surrounding water.

Some amino acids are small, and some are long and gangly. I didn't include size differences in my drawing, but size matters: big, bulky amino acids can't fit into small spaces. Nonpolar amino acids want to hide away inside the protein structure, away from water, but polar amino acids love the water. Negatively-charged amino acids adore positively-charged amino acids while avoiding other negative charges. When the string folds up, it does so in a shape that balances all these multitudes of interactions.

DNA gives directions for the order of amino acids.

In the nucleus, DNA is packaged up and stored on spools called histones, which organize those long strings of genetic material and keep them from getting all tangled up. DNA wraps around a core of eight histones, forming a small

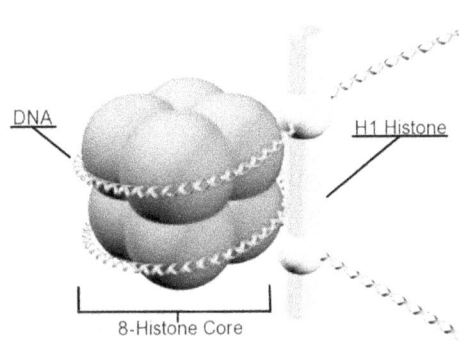

Figure 28: A 3-D representation of a nucleosome. Placed in the Public Domain by anonymous artist.

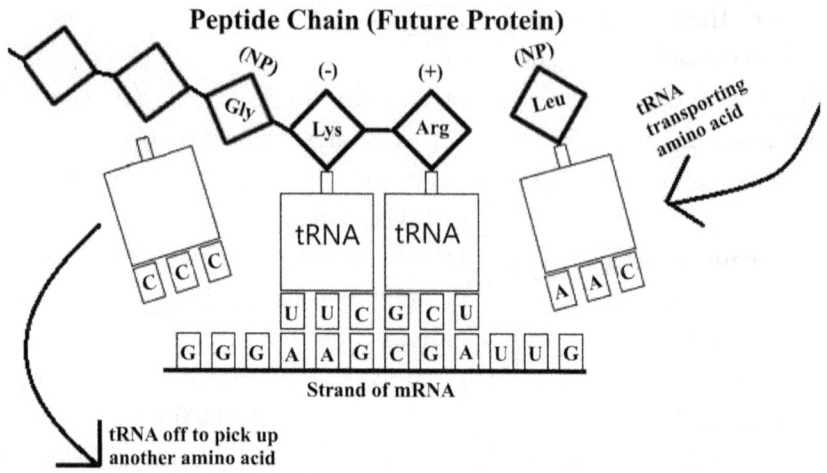

Figure 29: Protein synthesis. A polypeptide is being created by chaining amino acids together, according to an RNA template. Each amino acid is coded by a three-letter "codon." Positively charged, negatively charged, and nonpolar amino acids are indicated by (+), (-), and (NP), respectively.

package called a nucleosome (Fig. 28). To create a protein, the nucleosomes are shuffled around by enzymes (protein machines) until a specific stretch of DNA is exposed.

The DNA itself isn't chopped out and sent to do the job of making a protein. The original DNA needs to remain in the nucleosome, so a copy of the section is made. If the body needs a little bit of a protein, a few copies are made. If the body needs a *lot* of a particular protein, a large number of copies are made. The original DNA template is copied into a strand of messenger RNA (mRNA), which acts like a set of blueprints to go build a protein. It zooms out of the nucleus to find a ribosome - a protein factory that can read the blueprints.

Remember, it's all about stringing amino acids together in the right order. Once the mRNA reaches the ribosome protein factory, it feeds through the ribosome, where the amino acids are carted in on transfer RNA (tRNA) molecules. Each three-letter sequence of the mRNA (codon) matches up with the anticodon of a tRNA taxi cab, like in F above. C pairs up with G. A pairs with U. Boom boom boom, the amino acids are assembled in a row, attached one to the next, and spit out the other side of the ribosome in a chain.

Proteins cannot just form out in the cytosol - out there in the cell liquid. For one, amino acids give *off* a water molecule every time they link arms, and that won't happen in an environment already deluged with water. The peptide bonds are made in the ribosome factory, out of the wet. An amazingly efficient and wondrous dance takes place to produce proteins in our cells.

When I think about all the letters required to order the amino acids to make proteins that perform specific functions, my mind explodes and leaves little red patches of hemoglobin on the walls.

But, there's more! After leaving the ribosome, our young protein then heads through the carbohydrate factory of the golgi apparatus for additional bells and whistles, where it's packaged into a vesicle to get transported off somewhere. The new polypeptide is freed to go curl and bend and twist into its best conformation, after which it can go off to do its protein job. A protein's life gets busy.

To review:
1) Nucleosomes unwind to expose a section of DNA.
2) mRNA copies of that DNA are made.
3) The mRNA feeds through a ribosome where its translated into a chain of amino acids.
4) The chain goes to the golgi apparatus for bells and whistles.
5) The protein is shipped off to its jobsite.
6) The amino acid chain bends and twists into a protein machine (sometimes as part of a complex with dozens of other proteins).

By the way, the ribosome itself is a huge complex of (50-80) proteins. Which raises an issue. If the ribosome is required to build proteins, but the ribosome is made out of proteins, where did we get the ribosomes in our cells, back when we were cute little zygotes?

That's easy. They come ready-made in the egg cells that came from our mommies. The eggs come loaded with them, and mitochondria too, and all the other cell actors necessary for the embryo to develop.

So, if ribosomes are being made and passed down from mother

to child generation after generation, where did the original ribosome come from? Excellent question. In 2010, one George Fox offered a most likely order of events in the evolution of the ribosome.[2] Of course, he assumes it *had* to have evolved, he was not doing a feasibility study about whether it *could* evolve. We're about to see how improbable it is for random chance to form even one protein, let alone a giant protein complex.

Okay. Probabilities. Remember, the protein is made up of a string of amino acids, and each amino acid is coded for by a three letter "codon." There are 64 letter combinations possible for codons, and just 20 amino acids to code for. The amino acids methionine and tryptophan require just one specific three-letter code each. On the other hand, leucine and arginine can use any of six different codons, and most amino acids are called on by at least two.

	RNA Codons To Amino Acids Translation Table			
	A	C	U	G
A	AAU Asparagine AAC Asparagine AAA Lysine AAG Lysine	ACU Threonine ACC Threonine ACA Threonine ACG Threonine	AUU Isoleucine AUC Isoleucine AUA Isoleucine AUG Methionine	AGU Serine AGC Serine AGA Arginine AGG Arginine
C	CAU Histidine CAC Histidine CAA Glutamine CAG Glutamine	CCU Proline CCC Proline CCA Proline CCG Proline	CUU Leucine CUC Leucine CUA Leucine CUG Leucine	CGU Arginine CGC Arginine CGA Arginine CGG Arginine
U	UAU Tryosine UAC Tryosine UAA Ochre STOP UAG Amber STOP	UCU Serine UCC Serine UCA Serine UCG Serine	UUU Phenylalanine UUC Phenylalanine UUA Leucine UUG Leucine	UGU Cysteine UGC Cysteine UGA Opal STOP UGG Tryptophan
G	GAU Aspartic Acid GAC Aspartic Acid GAA Glutamic Acid GAG Glutamic Acid	GCU Alanine GCC Alanine GCA Alanine GCG Alanine	GUU Valine GUC Valine GUA Valine GUG Valine	GGU Glycine GGC Glycine GGA Glycine GGG Glycine

That's handy, because it makes plenty of single-letter mutations completely neutral. That is, they're neutral if there's a mutation in the third position. The first and second bases are a bit more crucial. Still,

---

2   Fox G.E. (2010). Origin and evolution of the ribosome. *Cold Spring Harbor Perspectives in Biology* 2(9):a003483, doi: 10.1101/cshperspect.a003483.

Appendix

whether the codon is CCU or CCC or CCA or CCG, the ribosome will still spit out Proline. There are even three "stop" codons.

This offers some flexibility for the code, which can be beneficial when copying mistakes are made and a C is swapped for a G on occasion. DNA is self-correcting and has an amazing process of double-checking copies for mistakes, but an occasional error slips through. Some are neutral, but some - like the one that causes sickle-cell anemia – can be deadly.

I suspect that most of the public overlooks how amazing DNA is. It's astonishing. The entire genetic code is found in each one of our cells, but different sections of DNA are shut off when they're not necessary for that cell. Heart cells know how to exploit only the sections of DNA that apply to them, and bone cells use only the parts that they need. From one set of instructions, our body produces the whole array of nerve and heart and pancreas cells. Not only that, but it produces them at the right timing. The cells follow directions on when to divide and grow and when to slow down or stop. Only relatively short sections of DNA are expressed at any one time, but each cell knows which portion of DNA needs to be exposed and read. That is a marvel.

Have you ever gone hiking, and you reach the top of a mountain only to realize that you've only crested a foothill? You lean over, panting. Then you stretch back up and gaze upon gorgeous mountain peak after mountain peak beyond and above you? That's how I feel about DNA. DNA has so many levels. The researchers started to understand the protein-coding regions, but the regulatory sections between the protein coding regions, the introns, are another set of mountain ranges we're just beginning to climb. Those introns are responsible for all kinds of regulation and timing, and when we throw in epigenetics - the biology of turning up the volume on certain kinds of genes - we find plenty of biological frontier land yet to explore.

When a string of mRNA is first produced, not all of it is for coding proteins. In the gene, the introns have to be cut out before the "exons" in the mRNA go off to the ribosome to do their protein assembly job. A giant RNA-slicer called a spliceosome takes care of this. The spliceosome is a collection of about 80 proteins and some nuclear RNAs, and it splices out the introns and connects the cut ends of the exons back into one string.

Here's the exceptionally amazing part of it all, the part that explodes like nitroglycerine in my mind. The very same pre-mRNA can be spliced in different ways, cutting off different exons for use in different proteins. Catch that? The same section can be changed up and used in different ways! In other words, parts of introns are sometimes made into exons in their own right. Ahhhh! I can't handle the complex choreography!

Also, did you notice that an 80-protein machine is used in making other proteins? This giant protein has to be assembled before it can go off and slice and splice stretches of DNA together to make other complex proteins. I can't grasp that.

This leads to one of the biggest chicken-and-egg questions in biology: which came first, DNA and RNA or the proteins they code for? Very large, complex proteins are used to transcribe DNA into mRNA and to then translate mRNA into additional proteins. DNA is required to make proteins, but proteins are required for the body to use DNA. Which came first? It's a paradox that biochemists have struggled over for decades.

They think RNA came first,[3] and maybe DNA first began replicating in viruses.[4] No, self-replicating proteins came first.[5] Wait, does that work? Wrenches. Wrenches everywhere.

It's good there are people with faith working on these things. I'm serious. They have full confidence that either RNA or self-replicating proteins came first - or something that acted as both - and they are working diligently to figure out how that could have happened. They have to do it, because I can't. I don't believe natural processes could ever have created this vast network of cell activity, so I can't

---

3   Robertson, M. P., & Joyce, G. F. (2012). The Origins of the RNA World. *Cold Spring Harbor Perspectives in Biology* 4(5):a003608, doi:10.1101/cshperspect.a003608.
4   Forterre P., Filée J., Myllykallio H. (2004). Origin and Evolution of DNA and DNA Replication Machineries. *The Genetic Code and the Origin of Life* (pp. 145-168). Boston, MA: Springer.
5   Guseva, E., Zuckermann, R, and Dill, K. (2017). Foldamer hypothesis for the growth and sequence differentiation of prebiotic polymers. *Proceedings of the National Academy of Sciences* 114 (36):E7460-E7468, doi:10.1073/pnas.1620179114.

be the one to fight that fight. I think those researchers are trying to explain how rain sculpted the Gettysburg Address on the inside of the Lincoln Memorial (and vastly, mountain-tops worse than that). But! But, I'm impressed by their dedication, and I'm happy to urge them on in their struggle for insight.

At this point, I want to give a shout out to a biochemist I admire. I've gone through Nick Lane's 2015 book *The Vital Question* about three times, and I want to express my deep respect for his work. Lane takes the evolutionary viewpoint in an honest, careful, straightforward way, and I am ardently impressed. He works to explain the origin of life through natural processes, and I think he does an admirable job of it. He boldly presents big problems, and he works to answer those problems. I think that's great, so great.

Lane makes a case for the origin of life having developed around alkaline hydrothermal vents - not hot black smokers, but calmer, friendlier hydrothermal vents. The vents would have provided a suitable environment and steady supply of materials required for early life, including natural proton gradients, H_2, CO_2, and metal sulfides.

Again, I'm skeptical, because I think the situation is far more complicated and difficult than naturalistic mechanisms can explain, but at least it's something folks can test. He can see whether anything interesting forms from bumping these building blocks of life underwater. Whether it works or not, his approach to answering the "vital question" is exactly what the evolutionary biologist ordered. Nick Lane tackles horribly difficult questions with boldness and creativity and honest care, and that's the right approach. Bully for you, Dr. Lane.

And if you read *The Vital Question*, then it's vital you also read the other side of the matter, which is Stephen C. Meyer's 2009 book *Signature in the Cell*, which offers an excellent, detailed look at the issue of origins and the evidence for engineering in nature. Meyer takes the Intelligent Design viewpoint in an honest, careful, straightforward way, and I am ardently impressed. They're both great books, and reading them back-to-back (or listening, because they're both available as audiobooks) is a good way to learn about the difficult origin of life question from both sides.

Okay then. How hard it is to make a light-sensitive spot?

3) A Simple Light-Sensitive Spot

Dr. S: You write so much. Does your hand cramp...?

AJ: The muscles are well used. They're-

Dr. S: Toned...

AJ: Yes. Although, my color sensors in my eyes got so tired the other night I stopped being able to see the color black. I only saw green... except on the periphery.

Dr. S: Which are those that see color, the rods or the cones?

AJ: The... I can't remember... I know Dr. Gurden knows. He could tell us.

Dr. S: You know, Amy Joy, Dr. Gurden is not the *only* professor who knows these things. I was asking you, but after that condescending answer-

AJ: HAH AHAHAHAHAH!! I'm just going to go crawl into a hole somewhere and --

Dr. S: I'm sure that you can find one.

AJ: --and pour sand on myself and stay there.

In *The Blind Watchmaker*, Richard Dawkins suggests that complex systems arose little by little, piece by piece, from simple to complex, because each step in the process improved chances for survival. That sounds great on the surface, but the problem is that there's no such thing as a small step in DNA programming. What appear to be small steps are giant cliffs.

Richard Dawkins suggests that the brilliance of the eye formed itself piece by piece, because possessing cells that detect light is better than nothing at all, and possessing blurry sight without a lens is better than no sight. Each step in the evolutionary development of the eye would have provided life with an adaptive advantage, and so Dawkins imagines that the eye developed slowly, step-by-step, because each improvement was better for survival than *not* having that improvement.

He calls this idea "cumulative selection." Beneficial attributes accumulate a little at a time, enabling a few light-sensitive cells to evolve into a complex eye with a lens and iris.

I see his point. I get where he's going. However, DNA stands as a roadblock against his idea that the eye can develop slowly. Why? Because each and every improvement in the eye only exists because of amazingly long, vast, complex lines of genetic code. He thinks that those codes can self-assemble after enough generations, but it's all imaginary on his part. I have no problem with his argument that having light sensitive cells is better for survival than being completely blind. That's not an issue. The issue is that light sensitive cells only exist because the DNA blueprints (and workers and project managers) were there first. Without DNA programming, the retina cannot place light-sensitive pigments into its rod cells.

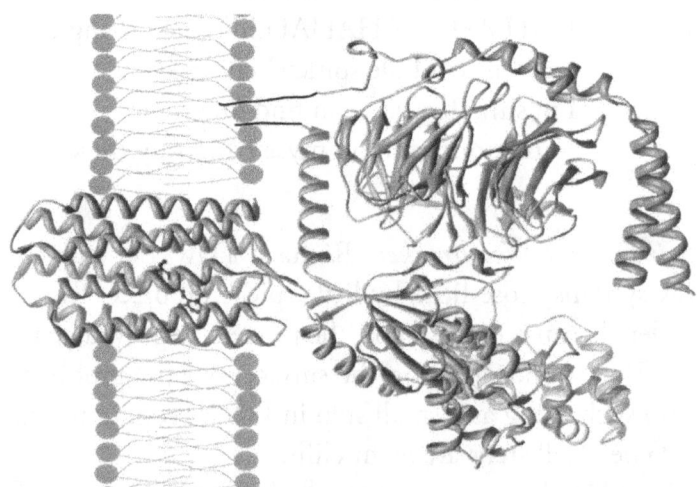

Figure 30: The seven alpha-helix trans-membrane structure of rhodopsin on the left, with the slender chromophore tucked inside (in black), ready to activate the three G-proteins of transducin on the right. Created using UCSF Chimera, compiled in Adobe Illustrator, and placed in the Public Domain by Dpryan at English Wikipedia.

It takes a lot to make a light-sensitive spot. A whole lot. It's not a small, simple step. It is an insurmountable 2000-foot cliff up to Observation Point, and worse.

All known animal eyes use opsin-based pigments for detecting light, including the starfish.[6] These opsins are all very similar in their structure, and rhodopsin is a common pigment used in animals for light detection in the photoreceptors at the back of the eye. Rhodopsin can be found in the retinas of cows and humans and chickens and fruit flies and all manner of creatures, and it serves the important purpose of converting low light into electrical signals that the brain interprets as sight. Rhodopsin is not responsible for seeing color, mind you.

That's the answer, Dr. Stillwell! Cones for color! Rods for light fishing. The rhodopsin in rods detects low levels of light.

In humans and cows and mice alike, rhodopsin consists of a single chain of 348 amino acids that form seven alpha helixes that wind back and forth across the cell membrane, with a chromophore tucked away inside the hydrophobic core of the protein (Fig. 30).[7] Throughout the animal kingdom, rhodopsin in other animals has that same structure – the same corkscrewing seven alpha helixes bending back and forth like a phone cord from the 1980s.

Those 348 amino acids form a chain that represents 1044 letters of DNA coding. The picture to the left shows three things: a cell membrane bilayer, rhodopsin, and transducin. The cell membrane looks like a column, and rhodopsin corkscrews seven times back and forth through it. Look carefully, and you can see the small, delicate black chromophore hiding among those corkscrews When light shines on the little retinal chromophore, it straightens up, which causes the rhodopsin protein to bend and change conformation, which activates transducin, seen with its three subunits, which causes a cascade of other events that boom-boom-boom send a signal to the brain that says, "light's on."

---

6   Garm, A., & Nilsson, D. E. (2014). Visual navigation in starfish: first evidence for the use of vision and eyes in starfish. Proceedings. Biological sciences 281(1777), 20133011, doi:10.1098/rspb.2013.3011
7   Franke, R., et al. (1992). Structure and Function in Rhodopsin. *The Journal of Biological Chemistry* 267(21): 14767-14774.

Rhodopsin is just the step in this whole process.

Now, there are some differences between the rhodopsins of humans and chickens and cows and fruit flies. Single mutations can lead to a variety of congenital retinal diseases, but rhodopsin can continue to function with a mutation or so.[8] However, rhodopsin only exists because DNA specifies the order of amino acids, so that they form those seven alpha helices that go back and forth across the cell membrane in the rods of the retina in a highly calibrated manner for each animal. The protein is shaped in such a manner that when light hits, the chromophore hidden away inside changes from an 11-*cis* to an all-*trans*-retinal, forcing rhodopsin to go through a series of physical changes.

The shape for Rhodopsin is coded for by 1044 DNA letters, and the next protein in the chain is even bigger.[9]

This is why it's so complicated. Those 1044 DNA letters that code for rhodopsin have to be queued up in a specific order to form the specific shape so that it will work. The words "California Condor" use only 16 letters and a space. Rhodopsin requires 1044 letters, and we all recognize that 1044 is more than 16.

I'm going to randomly type out 1044 characters using the letters A C T G, representing the nucleotide bases adenine, cytosine, thymine, and guanine, just for fun.

```
CGCGCGCGATATCGATGCGATGAACGCG
CGATCGAGTCGTATACGATGCATGCATGCATC
ATATCTCGCATATCTGATCGTATGCTGGATCT
AGGAGCGCGCCTCTTATCGCGGAGATCTCTA
GATCGGCATATGGTTTTTCCCCAGGCGAGAG
AGCTGATCGATAGTCATAGCGTAGCATATCGA
GAGCTCTAGAGCTCTTAGCATTCAGACGATC
TAAGGCGATCTAAGCGAAACCTCTAGAGCTA
TAGACTATACGATCGGAGAGATCTCTTATAG
```

8 Kazmin, R., et al.(2015). The Activation Pathway of Human Rhodopsin in Comparison to Bovine Rhodopsin. *The Journal of Biological Chemistry* 290:20117-20127.
9 Fong, S.L. (1992). Characterization of the Human Rod Transducin Alpha-Subunit Gene. *Nucleic Acids Research* 20(11):2865–2870, doi.org/10.1093/nar/20.11.2865.

Appendix

```
GCTACGATCGATAATCTCTAGGGATCTCTAT
ACGGATCAGATATAATATATATGGGATATATA
TCTCCCCAATTATTATCGGGCCTCGCTATA
TCTAGCTGAGGCTATAGCAAGGAGATCTCTA
GAAGCTTATATATCGCGCAATATCTCTTAAC
TCCATAACTTCTCTCTATAAGGATACGATAG
CATGACTATATATCTAAGAGCTATATATCTAT
CTATAGGCGAGACTATATCTAGGAGCATATTC
AGGAGATCTATTACTCTAGGAGACTATTATCT
CTCTATATCGAGAGATCTCTATAGAGGACTTA
TAGCGAAACCGGGCATCGTAGCTAGCCGAGA
GGACGCCCGCGCGATATCGTTATATCGATGA
TGCTGGTATATCGTGATACGTAGATCGTGTA
GCTAGTTAATCGTGATCTATGCTAGTCTATAG
CTGATGATCATATGGGCGCGCGCTCTCTCTC
TCTCTCTCATATACGATATTAATATATCGG
ATATCGGAGATATATAAAATCTCTTCTTAGAG
CTGATCGTAGAGCGTTCTCGAGATATCTGAG
AGGCATTTCGGGAATTTTAGGCGCGACTATT
CTATAGGCGAGACTATATCTAGGAGCATATTC
AGGAGATCTATTACTCTAGGAGACTATTATCT
CTCTATATCGAGAGATCTCTATAGAGGACTTA
TAGCGAAACCGGGCATCGTAGCTAGCCGAGA
GGACGCCCGCGCGATATCGTTATATCGATGA
CAATTATT
```

That's 1044 letters. That's a lot of letters to get into a specific order. The statistical probability of randomly pulling out any one of those four letters is $1/4^1$ - one in four. These four:

A C T G

The probability of pulling out any two letters in a specific order is $1/4^2$ - one in 16. These are the 16 possibilities:

AC	AT	AA	AG	CC	CT	CA	CG	TC	TT	TA	TG	GC	GT	GA	GG

The probability of pulling out *three* of those letters in a specific order is $1/4^3$ - one in 64. What follows are each of the 64 possibilities:

ACC	ACT	ACA	ACG	ATC	ATT	ATA	ATG	AAC	AAT	AAA	AAG	AGC	AGT	AGA	AGG
CCC	CCT	CCA	CCG	CTC	CTT	CTA	CTG	CAC	CAT	CAA	CAG	CGC	CGT	CGA	CGG
TCC	TCT	TCA	TCG	TTC	TTT	TTA	TTG	TAC	TAT	TAA	TAG	TGC	TGT	TGA	TGG
GCC	GCT	GCA	GCG	GTC	GTT	GTA	GTG	GAC	GAT	GAA	GAG	GGC	GGT	GGA	GGG

These happen to be the 64 possible codons that code for the 20 amino acids that make up proteins, including three possible "stop" codons.

The probability of pulling out four of those letters in a specific order is $1/4^4$ - one in 256. Getting 10 of those letters in a precise order is $1/4^{10}$, which calculates to 1 in 10^6, or one in a million. Already that's getting a little improbable, but not ridiculous. However, pulling out 1044 letters in order, using four letters, is 1 in 4^{1044} (aka 10^{628}) and that's what's termed "absurd." After all, there are only an estimated 10^{80} atoms in the universe.

Not any random sequence of 1044 nucleotide bases will do. The 1044 bases have to be in a specific order that tell the ribosome to chug out amino acids in a specific order that create seven alpha helixes that wind back and forth across the cell membrane like an old 1980s phone cord with just the right kinks in it.

Let's say DNA gets shuffling like cards in Vegas. Let's say there are 10,000,000 mutations per generation in the whole population of a group of organisms, and there are 10,000,000 generations per year. That's a ridiculously high number, but there's no need for me to be stingy in this thought experiment, because it would still take - statistically - 10^{614} years to randomly produce the DNA for rhodopsin.

Now! We can recognize that there are a variety of different forms of rhodopsin out there, in all manner of critters from fish to elephants, and there's always a minute chance that a working DNA sequence might have appeared serendipitously early. There's room for variation; most amino acids can be coded for by more than one codon. Let's say a working sequence conveniently appears early early on at only 10^{14} years. I've just cut off a vastly huge part of the statistically relevant amount of necessary time – 600 zeroes in a row – just to be crazy, unrealistically generous and account for all kinds of unexpected bits of good luck. But it's still not good enough. The entire universe is estimated at only 1.4×10^{10} years old, and 10^{14} years is still 10,000 times longer than the generally accepted age of the universe. It just doesn't look good from a straight statistical viewpoint, and wisdom says, "Go with the statistics."

Forming the rhodopsin protein isn't the only problem.

Let's say a rhodopsin wondrously forms in a cell, despite the probability against it. Let's say the gene is (randomly) transcribed and translated into a protein and produced abundantly, and there's a ton of the pigment. What value does it have? It's no good by itself. It has to be utilized as part of a system. Rhodopsins work because they change shape in response to light, which starts a cascade of events that eventually communicates to a part of the brain, where those signals are interpreted as "light." If there's no relay system and no optical part of the brain, it doesn't matter whether rhodopsin's hanging out or not.

Cumulative selection can only work on DNA sequences that provide an adaptive advantage by producing functional proteins in the right place at the right time of development. What did the unfinished gene for rhodopsin do for countless generations while it hung around being useless until the system around it had mutated sufficiently to start doing the job of creating electrical signals that the brain interpreted as light? This scenario requires cells out there producing random proteins for no reason. Cells filled with useless proteins. Produced at the wrong time and in the wrong place for no reason. It's only when rhodopsin actually *works* that it can offer an advantage to help a critter survive a bit better than those who can't detect light.

Until then, though, there's no cumulative selection. There's no advantage. There's nothing pushing the DNA in one direction or another.

Also! Remember that DNA is wrapped up around histone proteins in nucleosomes. There are about 6 billion base pairs of DNA wrapped up in the 46 chromosomes in each human cell. The cell has to be able to access specific nucleosomes with specific base pair sequences at certain times. It's all very ordered. My room is not that ordered. My desk is absolutely not that ordered, and here I am an intelligent being capable of creating organization.

Excuse me, Mr. DNA. Please organize your base pairs into the correct sequence to code for rhodopsin, and also to code for all the subunits of transducin, and wrap yourself up around a number of histones so all that the DNA doesn't get all tangled... but... be available for transcription when required. Could you please do that? All by yourself?

Oh! And once transducin is activated, setting off the rest of the phototransduction cascade of events, could you set it up so that some other proteins naturally *stop* its activation, so that everything can reverse and return to normal, because I don't want to keep seeing explosions of light while I'm trying to sleep? Could you do this through random processes that are primarily destructive?

In other words. Something so "simple" as detecting light (and this is just the purple-red wavelengths of light) requires proteins in specific conformations interacting with other proteins in specific conformations, which form the shapes they do because they have a certain amino acid sequence, which is coded for by thousands of DNA letters in a specific order. And those proteins connect to a nervous system that is ready to interpret the electrical signals produced by these changing protein conformations.

A light-sensitive cell is not a small step. It's a trip to the Moon. In all honesty, with all humility, I'm floored by the whole thing. This is just one sliver of the processes required for sight, not to mention all the other multitude of whirring mechanisms we find in living systems, and it's ridiculous to argue that it all programmed itself.

There's another side to this too. Tracing any evolutionary relationship from animals with simple eyes to those with complex eyes is pretty much an exercise in the absurd. Octopuses have a lens and iris like we have, but not because they're closely related to us. In fact, they can be considered superior to us in some ways, because octopuses and squids have no blind spot like we have (because the optic nerve of us vertebrates blocks light). The world is filled with "convergent evolution," in which different organisms possess similar forms and functions completely independently of each other.

And then there are the trilobites. Trilobites, which existed in the Cambrian (at the bottom of the geologic column) already had fully formed compound eyes, complete with calcite lenses and color vision. Right down there at the beginning of recorded multicellular life, fully-formed complex eyes already existed. Dawkins can imagine the slow development of eyes over time, but what we find – the evidence we have in reality – is that complex eyes already existed at the beginning of the Paleozoic.

Dr. Stillwell has reminded me of the vast amount of time, the depth of time available in the geologists' 4.6-ish billions of years of earth's history, the unfathomable amount of time available there. Yet, the cell itself doesn't have billions of years. It has a few minutes or days or weeks, and the natural forces of our world are destructive forces. The only reason that life reproduces and builds itself is that it has a program written to make order out of chaos.

Again, I don't think the I.D. theorists are making a "God of the gaps" fallacy when they say, "An Intelligent Designer did it." I think they are the only ones being practical. They're the only ones facing the facts, because the only time *we see* code in the real world is when an intelligent mind writes code. It's the only explanation that comes even close to working.

We want to understand the nuts and bolts of biological processes, and the universe is set up for natural processes to continue on and on, reproducing life until The End. However, the biological disciplines do not have solid answers about how life formed. And everywhere we look, we see astounding feats of engineering and chicken-and-egg paradoxes. I think that the folks who *refuse* to consider an Intelligent Design explanation are the ones with a philosophical problem.

They've made rules for themselves, stuck themselves inside a box of their own creation, and won't accept any possibilities outside of that box.

I suggest that if evolution creates functional machines from scratch, we should be able to determine the cause and effect mechanisms. Evolution had to have written the codes for genes a multitude of times throughout history, and so making order from disorder should be the normal trend of our universe. We should be able to model and map out the series of events that made wonders like the eye, and all the pieces should eventually fit together about how that happened. After all, if Evolution created us all, then there are countless cases of super amazing biological machines self-producing. If that's how the universe really works, it shouldn't be too hard to figure it out, because those marvels have taken place repeatedly for four billion years.

If, however, God created animals and plants fully formed at the beginning, then the biological world should be filled with paradoxes that offer more problems than answers. There should be holes all over the place, with unexpected, unexplainable *non sequiturs*. I started off skeptical that Evolution could do the trick of building anything, and I'm still a skeptic.

Appendix

Figure 31: "Earthrise" taken by astronaut Bill Anders during the Apollo 8 mission on December 24, 1968. Credit: NASA.

4) Entropy

During the Apollo 15 mission in 1971, Jim Irwin became the 8th person to walk on the moon. The warmth and life of Earth in all the blackness of space struck Irwin, and his friend Bob Cornuke recalls a conversation in which Irwin commented that, "Everything about space wants to kill, wants to rob you of life. It's an unsympathetic and hostile place."[10] This is a common sentiment among those who work on space missions, that Earth is a beautiful blue oasis of life in a universe bent on destroying us.

I keep mentioning that the random processes of our world are mostly destructive. I keep touching on it, but I shouldn't assume everybody understands why.

The Second Law of Thermodynamics states that entropy (randomness) always increases in a closed system. Heat constantly moves from hotter objects to colder objects. Your coffee will give up its comforting warmth to the surrounding air of your office if you don't enjoy it first. Entropy is the measure of disorder in a system, and that disorder is a ravenous, scattering, punishing force. A chew-happy dog in a room of nice furniture.

Randomness always increases in the end. We can depend on it. In this world, this universe, randomness beats things to pieces. It's

---

10   Cornuke, R. (2018) *Tradition*. Coeur d'Alene: Koinonia House.

the law, and we'll never get a perpetual motion machine because of it. Some energy is always lost as heat, making it unavailable for work to keep the machine going. Chug chug chug, it slows and dies.

Are there periods of increased organization in our world? Absolutely. Our bodies run programs that can take energy from food to do the work of building cells and directing processes, but randomness noses in all the while. Incessantly. (Go! Go lie down!) It never really does stop. Every cell in our body is fighting it, and eventually they'll give in and die. We build machines from steel, but they rust and break down. Our cars, our houses, our refrigerators and socks all wear out and fall apart. That's the natural trend in our universe, the one universe we can observe. It's a net loss of order.

If a little protein molecule wanted to self-form out there on the young Earth, all the violent forces of our universe would bust it up before it could get anywhere. That's the reality. Ocean waves might stack up driftwood, but they don't build a cabin when they do it. If they managed to build something close to a cabin, the next storm would tear it to bits. The random House always wins. (Ha.)

Final conclusion? It's bad form to ignore Entropy when formulating biological theories. I don't recommend.

5) OF MICE AND MEN

This brings us back to the issue of frog kidneys and chloroplast DNA. This is where the whole thing gets tangled up worse than a cat in grandma's yarn collection.

Frogs and horses and birds all have lungs and eyes and skulls. If we examine the limbs of frogs and horses and birds, we see the same basic bone structures. The bones in the wing of a bat can be easily compared to the bones in our hands. These similarities are called homologous structures. The existence of rhodopsin might be a miracle, but we still find that human and bovine and mice rhodopsins are remarkably alike.

How to handle that? I've been trying to wrap my head around it, and my head isn't that flexible.

There are two different ways to look at this. They are both easy to say, but it's difficult to track down which works best:

1) Homologous structures are evidence of evolutionary relationships and strongly support the evolution of all creatures from common ancestors.
2) Homologous structures are not evidence for evolution. They are the results of design by a designer who created those structures for similar functions.

The first isn't difficult to understand. We have hearts, and chickens have hearts, and fruit flies have hearts. It makes sense that we would have similar structures because we share ancestors, and natural selection acting on mutations has created the variety we see today. That makes initial sense. Getting the hearts into humans and chickens and fruitflies in the first place is an issue, but we get the idea.

The second also makes sense. Tables have legs and chairs have legs and ladders have legs because we need them to have legs to do their jobs. They are similar, because we build them that way. Machine screws and wood screws and drywall screws all have similar shapes, with threading, but they were intentionally made with important differences because they were intended for different niches in the construction of buildings.

Protein shape determines protein function, which means that any creatures that require certain proteins will have to have relatively identical stretches of DNA. Right?

Everybody on all sides can agree that chloroplast DNA is similar in all land plants because it's what *works* to make photosynthesis take place in the cell - whether it was designed on purpose to keep plants alive, or whether it works and that's why it keeps plants alive. The data can fit either model.

That's the huge puzzle, the question we all need answered well: **when are similar DNA sequences the result of common ancestry, and when are they not?** If the same lines of code appear in fruit flies and in mice, does that mean the code is extremely old and has been "conserved" since their common ancestor, or does it mean they have the same code writer behind their programming?

Mice and humans share about 85% of the protein-coding regions of their genome. Some portions are 99% identical and others are 60% identical, but the fact that they are so close is extremely handy for research on human diseases. We can tinker with mouse orthologs, - the mouse versions of a gene - and study how the tinkering affects our fuzzy friend and thus how it affects us.

Figure 32 shows a comparison of the mouse and human genomes. A large number of genes from the 20 mouse chromosomes (on the right) are found somewhere in the human genome (on the left). For the most part, they're not in the same order, but researchers are able to find the multitude of genetic homologs. "Ahah! There's rhodopsin in the mouse genetic code! I knew we'd find it somewhere."

Human and murine (mouse) rhodopsins are almost identical. Out of the 348 amino acids in rhodopsin, there are 18 amino acid residues different between humans and mice.[11] These variations slightly change the shape of the rhodopsin, tweaked in mice to work

---

11   Whited, A. M., & Park, P. S. (2014). Nanodomain organization of rhodopsin in native human and murine rod outer segment disc membranes. *Biochimica et Biophysica Acta* 1848(1 Pt A): 26-34.

Appendix

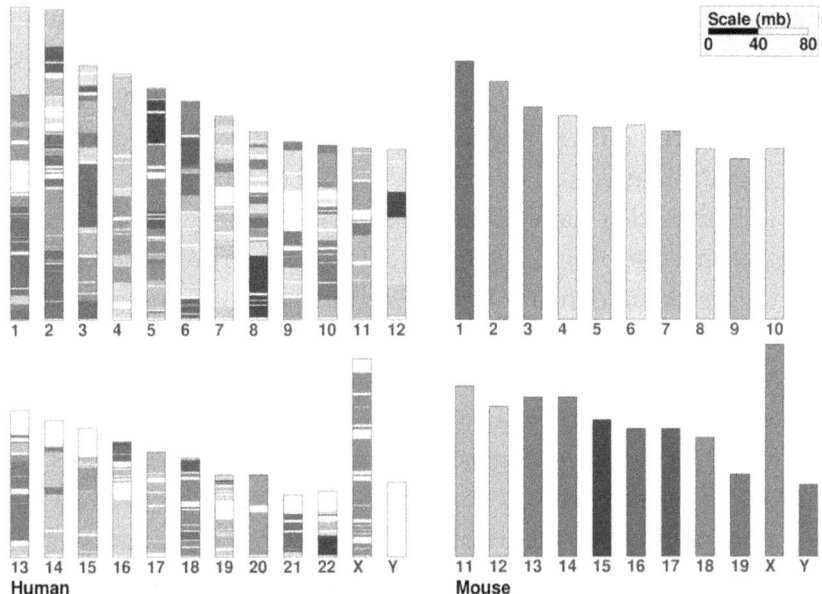

Figure 32: Visualization of synteny between human and mouse genomes. Mice have many of the same genes as humans, even though they are often not found in the same locations or even on the same chromosomes. Source: BMC Bioinformatics. *2007 Mar 8;8:82 [PMID: 17343765], Authors A U.Sinha and J. Meller.*

for mice, and tweaked in humans to work for humans, but they still function the same way.

Again, it is exceedingly handy that human and mice genes are so similar. For instance, we can make purposeful changes to the murine rhodopsin gene in order to better understand congenital human night blindness.

That's all true and just. However! Does that mean humans and mice evolved from a common ancestor? Is it actually feasible for mutations to have led from a common ancestor to the humans on one side and the mice on the other?

There are a lot of mutations to be found in human rhodopsin. Mutations have littered the human gene pool, after all. A letter change here. A letter change there. Entropy at work. There are at least 100 different mutations at 60 different locations in the human rhodopsin gene linked to *Retinitis pigmentosa*, a degenerative disorder that leads to reduced night vision and peripheral vision.[12] Many of these mutations produce single-amino acid differences, but they can lead to improper folding or transport problems or other troubles that cause the protein not to function correctly. Single amino acid differences matter.

On the other hand, if one amino acid change can lead to the degeneration of vision in humans, then it doesn't seem the shape of rhodopsin is particularly flexible within a species - not if we need it to work. This makes the 18 amino acids of difference between humans and mice a big deal. Any one of those 18 changes could potentially break the protein.

This is one thing that bothers me. If Evolution made us, and mice and humans had a common ancestor, then there were a large number of in-between critters. As those amino acid changes wriggled in, they still had to work for each one of those in-between critters. And not just for rhodopsin, but for all the proteins.

Let me repeat that. It's not just one gene that is changing. It's every gene, *all* the proteins that exhibit differences between humans and mice. As the genes changed, they still had to make a functional creature capable of surviving long enough to produce offspring.

Are the transitional forms even feasible? We should make models to check and see. Remember, we don't have actual common ancestors or in-betweens to point to, to look at, to snag DNA from and analyze, and we have scant fossil record evidence of transitional forms. We just have our imaginations.

Is it a big deal that mice and humans have a genome that is 85% identical in the protein coding regions? Yes! Does that suggest a common ancestor? Sure! But, it doesn't necessitate one. What transitional forms would actually work? That's the question.

Somebody out there is shouting, "There are jumping genes! Chromosomes and genes can move around. Mutations do exist without causing death, and they give us a huge amount of variety! We see what we should expect to see after millions upon millions

---

12 Kazmin, R., et al. (2015).The activation pathway of human rhodopsin in comparison to bovine rhodopsin. *The Journal of Biological Chemistry* 290(33):20117-20127.

of years of divergence. Thats what we see with mice and humans, millions of years of mixing up genes from their common ancestor."

I understand, but did it work that way in reality? We see bushy trees with all kinds of horse branches and bushy trees with all kinds of rhinoceros branches. But I don't see the ancient critters that connect horses to rhinos. It appears the branches are stretching, stretching and mixing-it-up from the original horses or rhinos in the center, and we don't know whether in-betweens ever existed.

OF WOMEN AND MEN

On the other hand, human beings are 99.9% genetically identical to each other, but there are a multitude of mutations surviving in our gene pool. There are 3 billion base-pairs in the 23 haploid chromosomes of the human genome, 6×10^9 in all 46 chromosomes. In 2015, the 1000 Genomes Project told us that the typical human genome varies from a reference genome at 4.1 million to 5.0 million sites, 99.9% of which are single letter mutations.[13] Differences have crept in over the years, and sometimes those cause blue eyes and freckles and sometimes they cause cowlicks or different-shaped ear lobes, and sometimes they cause serious genetic diseases.

So many mutations have crept in that we can no longer marry our first cousins, lest we compound our family's genetic mistakes. Each of us has potentially deadly mutations from one parent or another, but if our parents aren't closely related, they fill in the gaps for each other. We might get a broken gene from one parent but still get the healthy version from the other. Good thing we have two sets of chromosomes; every gene has a backup.

This backup system allows certain destructive genes to float around in the gene pool without killing all their hosts. They even occasionally help people. That's why sickle cell anemia is such a big deal to evolutionary biologists. Having two versions of the bad gene causes sufferers to die painfully from sickle cell disease, but having one bad gene means only half the blood cells are sickle shaped,

---

13 1000 Genomes Project Consortium, Auton, A., et al. (2015). A global reference for human genetic variation. *Nature* 526(7571): 68-74.

which helps folks survive malaria. Is this tolerance for broken genes sufficient to allow the evolution of mice and humans from a common ancestor, with the in-betweens? Or. Does it just help organisms survive entropy a little longer?

Still, it's clear that we're all related and originally diverged from a small group of people.

In 1987, mitochondrial DNA pointed to a single mother for all modern humans, dubbed Mitochondrial Eve.[14] The genomes of people across the world have since been sequenced, and analysts have been able to paint a pretty good picture of how human populations migrated from northern Africa and the Middle East into all parts of the world.[15]

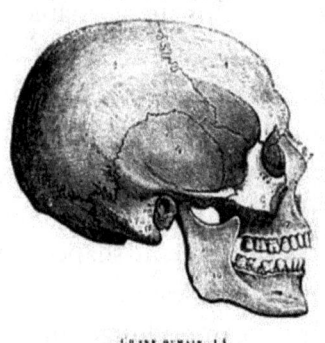

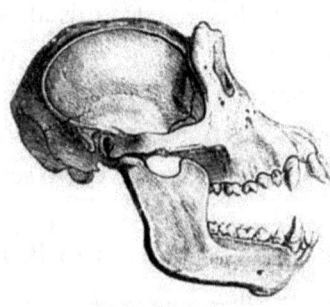

Figure 33: Human v. chimpanzee skulls. By Paul Gervais from Histoire Naturelle des Mammifères *(1854).*

Right. So, what about apes?

Of Chimps and Men

Did humans and apes descend from a common ancestor? That's the normal evolutionary assumption.

We do have a high genetic correlation with apes. We've heard for decades that we are 98-99% similar to chimpanzees. But that's not quite right.

A 2004 comparison of human and chimpanzee chromosome 21 showed single-base substitutions in 1.44% of the chromosome "in addition to nearly 68,000 insertions or deletions."[16] That was just one chromosome. The 2005 sequencing of the entire chimpanzee genome

---

14 Cann, R., Stoneking, M. & Wilson, A. (1987). Mitochondrial DNA and human evolution. *Nature* 325: 31–36, doi:10.1038/325031a0
15 Mallick, S., Li, H., Lipson, M. et al. (2016). The Simons Genome Diversity Project: 300 genomes from 142 diverse populations. *Nature* 538: 201–206, doi:10.1038/nature18964
16 Watanabe, H. et al. (2004). DNA sequence and comparative analysis of chimpanzee chromosome 22. *Nature* 429: 382–388.

told us there are in total about 35 million "single-nucleotide changes, five million insertion/deletion events, and various chromosomal rearrangements" when comparing human and chimpanzee genomes.[17] Five million insertion/deletions represent just 0.16% of the genome (if they are all single letters).

But, there's more to to it.

First, we have to be wary of the earlier chimp-human comparisons, because of the possibility of human DNA getting into the mix of ape DNA. Human DNA has been found to persistently contaminate non-primate genome databases,[18] which suggests that primate databases might also be contaminated. So, that's an issue. The 2018 chimp genome assembly (panTro6) was created without using a human genome as a scaffold, which is good, because it lessens the threat of contamination.[19] (Thanks to geneticist Jeffrey P Tomkins for his many articles analyzing the issues involved with human and chimpanzee genomes.)

Even then, it's important to read the fine details of genome comparisons. A 2018 study aligned the chimp panTro4 genome assembly to the human hg38 genome, and the researchers reported a similarity of "98% identical across 96%" of the genome's length.[20] That sounds great. It sounds like we're 98% identical to chimps. This summary statement gives a layman the impression that what makes humans unique is found in the 1.93% of the single nucleotide base differences between humans and chimpanzees.

No, there's more. That 98% only represents the parts of each genome that *could be* aligned against each other. The report also explains that, "Approximately 306 Mb (9.91%) of the human sequence did not align to the chimpanzee sequence, while 138 Mb (4.15%) of the chimpanzee sequence did not align to human."[21]

---

17 Chimpanzee Sequencing; Analysis Consortium (2005). Initial sequence of the chimpanzee genome and comparison with the human genome. *Nature* 437(7055): 69–87, doi:10.1038/nature04072.
18 Kryukov, K. & Imanishi, T. (2016). Human Contamination in Public Genome Assemblies. *PLoS One* 11(9): e0162424.
19 Kronenberg, Z.N. et al. (2018). High-Resolution Comparative Analysis of Great Ape Genomes. *Science* 360 (6393):eaar6343.
20 Marçais, G., et al. (2018). MUMmer4: A fast and versatile genome alignment system. *PLoS Computational Biology* 14(1): e1005944, https://doi.org/10.1371/journal.pcbi.1005944.
21 *Ibid.*

Wait. Wait then. So, it's not the whole genome that's nearly identical? That means we have 14.06% between the two species that can't be lined up for comparison? With the 1.93% single nucleotide polymorphisms (SNPs), it's closer to an 84% base-for-base match. I mean, that's a whole lot of identical bases, but it's not 98%.

Richard Buggs, Professor of Evolutionary Genomics, Queen Mary University of London, checked out the genomes himself. He compared the hg38 and panTro6 alignments from the UCSC genomics website and reported the following.

4.06% had no alignment to the chimp assembly

5.18% was in CNVs relative to chimp

1.12% differed due to SNPs in the one-to-one best aligned regions

0.28% differed due to indels within the one-to-one best aligned regions

The percentage of nucleotides in the human genome with one-to-one exact matches in the chimpanzee genome was 84.38%[22]

Buggs only counted the lack of alignment from human to chimp (4.06% no alignment + 5.18% in CNVs) and not the chimp to human, but the final take-away was that the chimp genome exactly matched the human genome just over 84% of the time. (CNVs are copy number variations, places within the genome where the same sequence is repeated multiple times. SNPs are single nucleotide polymorphisms - single letter differences. Indels are insertions and deletions. All the differences took their toll.)

Those base-for-base matches do give the impression that humans and chimps are related. There's a lot of similarity, but the differences are vital. We humans are self-aware beings who create musical instruments and build bridges and space shuttles, while chimps are

---

22   Buggs, R. (2018, July 14). How similar are human and chimpanzee genomes? Retrieved December 8, 2019, from http://richardbuggs.com/index.php/2018/07/14/how-similar-are-human-and-chimpanzee-genomes/.

muscular animals that throw poop.

When Buggs analyzed the panTro4 assembly, he found a one-to-one match only 82.34% of the time. The panTro6 assembly is higher quality all around, so we can stick with that one, but the bottom line is that 84% is not 98%.

What about Neanderthals? Neanderthal DNA has been sequenced as well, and most Europeans have some Neanderthal genes to this day. All indications suggest the Neanderthals were fully human but came from a line that died out. They buried their dead[23] and made musical instruments[24] and used cutlery on their rock doves.[25] They did things only humans do. I won't do it now, but I do want to focus on extinct hominins in a future volume.

OF PIGS AND MEN

We're also exceptionally similar to the pig! In fact, pigs are seen as better biomedical models than mice in some ways, because of similar biochemical and physiological functions to humans. One of the biggest surprises about pigs (the family Suidae) is that they have transposable elements called SINES (short interspersed elements) that appear to have evolved from 7SL RNA like humans and apes. Most SINES are believed to have evolved from tRNAs, so this makes pigs kinda special. It suggests that they are more closely related to humans than most other placental mammals from the supercontinent of Laurasia.

> The similarity of PRE-1 and Alu elements further revealed a hidden kinship, such that suidaes are more closely related to the suraprimate/primate than any other laurasiatherias based on the 7SL RNA derived SINE composition of their genomes.[26]

---

23 Wong, K. (2014). Ancient Burial. *Scientific American*, 310(3): 19.
24 A famous "Neanderthal flute" was found in 1995 in the cave of Divje babe I in western Slovenia
25 Blasco, R. et al. (2014). The earliest pigeon fanciers. *Scientific Reports*, 4(5971): 1-7, doi:10.1038/srep05971
26 Yu, H. (2015). Genome-wide characterization of PRE-1 reveals a hidden evolutionary relationship between suidae and primates. *BioRxiv*, doi: https://doi.org/10.1101/025791.

Okay okay okay. So, we can compare human and mice and pig and cow and ape genomes and easily say, "We're the same. We all descended from a common ancestor."

Except. The differences are huge. Millions upon millions of DNA base differences, and the in-betweens matter. The transitional forms would have had to survive, but we're dreadfully shy on transitional forms to study.

What common ancestor did chimps share with pigs that they both have SINES? How long ago, and where are all the pig-chimp variations?

Convergent Evolution

There's another issue in all of this. Similarity doesn't necessarily require biological relationship! There's convergent evolution too. We see the same sorts of things allegedly evolving separately in different family groups on different continents.

In other words, the same body part or function appears in critters not considered closely related at all. For instance, porcupines, hedgehogs, and echidnas all have protective spines, even though they're completely different sorts of creatures. Echidnas are Australian monotremes (which are neither marsupials nor placentals) like the platypus. Porcupines are rodents, but hedgehogs are more closely related to shrews (which are not rodents).

Remember, the not-so-stupid appendix that turns out to be useful after all.[27] The appendix has evolved independently at least 32 times.[28] Having an appendix doesn't mean that creatures are related! It just randomly sprouts up here and there.

Notice all the creatures that can fly? There are flying bugs and birds and mammals and dinosaurs. There are swimming bugs and birds and mammals and dinosaurs. There are hopping birds and bugs and mammals and dinosaurs. Amphibians and reptiles get around

---

27 Sanders, N. (2013). Appendectomy and Clostridium difficile colitis: Relationships revealed by clinical observations and immunology. *World Journal of Gastroenterology* 5607–5614.

28 Smith, H., Parker, W., Kotzé, S., & Laurin, M. (2013). Multiple independent appearances of the cecal appendix in mammalian evolution and an investigation of related ecological and anatomical factors. *Comptes Rendus Palevol* 12(6): 339–354.

too. The general explanation is that they all fill niches, but that's not a real explanation. It's just an observation. It doesn't explain the code involved in forming all these things. The DNA coding that goes into dragonfly wings is nothing like the coding that goes into sparrow wings. Yet. They both have wings.

Do wings just evolve easily after all? H'yeah, lots of problems with that idea.

OF FRUIT FLIES AND MEN

What about the fruit fly? How closely are we related to the fruit fly? *Drosophila melanogaster*, the geneticist's best friend, also has eyes and a heart and has sex to produce babies. They have only eight chromosomes to our 46, but according to Sharmila Bhattacharya of NASA's Ames Research Center, "50% of fly protein sequences have mammalian analogues."[29] That doesn't mean their genome aligns 50% with ours, but they do use the same proteins that we do. And despite the vast genetic differences, they still have a lot of the same parts! They have faces and eyes! Fruit flies have sex like mammals have sex, with all the necessary junk involved. And by the way, if male fruit flies are sexually deprived, they're more likely to be interested in alcohol.[30] Ha! Wildly different DNA, but same parts.

One of those puzzles that muddles the whole issue for me is the complexity of fruit fly rhodopsins. No, this is good! Check this out!

Fruit flies don't have just one type of rhodopsin. They have *seven*. Their rhodopsins are still proteins made of seven helices winding back and forth through the cell membrane, but there are differences that allow fruit flies to detect various wavelengths of blue and green and ultraviolet light. Our rhodopsin only detects red-purple wavelengths of light. Fruit fly light detection is more sophisticated than ours in this way.

But it gets weirder. When comparing the amino acid sequences between the seven rhodopsins used by fruit flies, we find that they

---

29 NASA (Feb 3, 2004). The Fruit Fly in You. *NASA Science*.
30 Shohat-Ophir, G., et al. (2012). Sexual deprivation increases ethanol intake in Drosophila. *Science*, 335 (6074): 1351–1355, doi: 10.1126/science.1215932.

are only 27%-73% alike.[31] It's not as though those rhodopsins are almost identical to each other (in one little critter). They're not. Their amino acid sequences are widely different, and yet, they all make the same basic shape. That means at least two things. 1) Widely different amino acid sequences for related proteins don't require a grossly distant genetic relationship. 2) Detection of different wavelengths of light requires precise and unique rhodopsin specifications! That sounds like calibration to me.

Seven rhodopsins. To recognize different light wavelengths. In a common little fruit fly.

Junk DNA

Does evolutionary theory offer solid predictions about genes? In 1972, renowned evolutionary biologist Susumu Ohno predicted that most of DNA would be useless, leftover "junk" DNA from millions of years of body-building and tossing away old info. When the genome was sequenced, and vast areas of DNA were found that didn't code for proteins, folks thought Ohno was right.

In 1998, atheist Richard Dawkins stated in *The Skeptic*:

> Genomes are littered with nonfunctional pseudogenes, faulty duplicates of functional genes that do nothing... And there's lots more DNA that doesn't even deserve the name pseudogene...It consists of multiple copies of junk, "tandem repeats", and other nonsense which may be useful for forensic detectives but which doesn't seem to be used in the body itself. Once again, creationists might spend some earnest time speculating on why the Creator should bother to litter genomes with untranslated pseudogenes and junk tandem repeat DNA.[32]

Even back then, I suspected that what they *thought* was junk

---

31 Senthilan, P. R., & Helfrich-Förster, C. (2016). Rhodopsin 7-The unusual Rhodopsin in Drosophila. *PeerJ*, 4: e2427, doi:10.7717/peerj.2427.
32 Dawkins, R. (1998). The Information Challenge. *The Skeptic* 18(4): 21-25.

would ended up having value. I wasn't alone; the Intelligent Design guys all agreed. We didn't know the purpose of junk DNA yet, but we assumed that it had all been designed on purpose. In making his case for the scientific value of taking a design approach to nature, William Dembski wrote in 1998:

> Thus on an evolutionary view we expect a lot of useless DNA. If, on the other hand, organisms are designed, we expect DNA, as much as possible, to exhibit function. And indeed, the most recent findings suggest that designating DNA as "junk" merely cloaks our current lack of knowledge about function.[33]

Then! In 2012, the journal *Science* declared that all those non-protein coding regions had value after all, with an article entitled, "ENCODE Project Writes Eulogy for Junk DNA."[34] After a decade and a consortium of researchers exploring those genetic oceans, the Encyclopedia of DNA Elements (ENCODE) revealed vast treasures of information. All that previously mysterious DNA worked to benefit the body after all. It included the regulatory parts of DNA, the directors and time keepers and project managers of Cell World. Studying just 147 kinds of cells among the human body's thousands of cell types, they found 80% of the DNA sequences accounted for, with hope they'd figure out the final 20% in time.

The genetic code is not islands of genes in an ocean of garbage. Some of it codes for proteins, yes, but other parts are responsible for DNA duplication and repair, cellular stress response and forming cellular structures, nucleation centers, cell division, immune response, memory, fear responses, metabolism and on and on. Protein-coding genes are merely one part of the vast number of things that DNA handles.

Take note. Richard Dawkins was incorrect and the I.D. guys were correct in their views regarding junk DNA.

---

33 Dembski, W. (1998). Science and Design. *First Things*, https://firstthings.com/science-and-design/.
34 Pennisi, E. (2012). ENCODE Project Writes Eulogy for Junk DNA. *Science* 337(6099):1159-1161, DOI: 10.1126/science.337.6099.115.

Orphan Genes

Then, there are "orphan" genes that have no relatives. On one hand, animals that aren't closely related can have the same parts. On the other hand, unique genes appear out nowhere, with no family history.

Orphan genes are sections of DNA found only in one species or genus, with no homologs of the gene detectable in other supposedly close relatives. The more genomes that are sequenced, the more of these introverted, lonely genes are found! True orphan genes aren't shared with anybody else, and all critters seem to have them. Evolutionary biologists predicted orphan genes to be rare or nonexistant, "so improbable as to be impossible for more recent evolution."[35] Yet as they appear increasingly common, they've begged an explanation.

The general approach is that these *de novo* genes appear because of very rapid evolutionary duplication or divergence in the regions of the genetic code that didn't previously code for proteins.[36] In other words, areas once considered junk DNA are now seen as the nurseries of the protein world, with lots of room for protein-coding genes to develop. New proteins are then believed to be adopted by the body and ushered into cellular processes.

Again, that all seems improbable. Do orphan genes exist? Yes. Does that mean they *had* to have evolved somehow, because that's the only option? No. Some orphan genes are highly conserved within the species and appear necessary for that species' livelihood. That is, the critter wouldn't have survived without that gene before it was developed into its current state.

Orphan genes present plenty of other problems. Genes have to be located and exposed according to carefully directed timing. It isn't simply a matter of forming new genes out of stretches of (allegedly)

---

35 McLysaght, A & Guerzoni, D. (2015). New genes from non-coding sequence: the role of de novo protein-coding genes in eukaryotic evolutionary innovation. *Philosophical Transactions: Biological Sciences* 370 (1678): 1-8.

36 Tautz, D. & Domazet-Loso, T. (2011). The Evolutionary Origin of Orphan Genes. *Nature Reviews Genetics* 12:692-702.

unnecessary introns; the body has to know how and when to use new genes. Who tells them how to do their job if their job never existed before?

The cell is highly choreographed. Proteins have to have addresses on them. They are ordered up by the body because the body needs them, and they are carted off to do their necessary work. Any rogue new proteins would be picked up by a lysosome garbage truck and destroyed. That's reality.

It's easy to say, "Oh, this codfish has antifreeze genes that its relatives don't have. It must have built those genes."[37] If the genes were already in the gene pool, and the fish simply duplicated those antifreeze genes to endure colder and colder water, then that's one thing. However, if the codfish has genes that no other fish have, that's another. And some organisms have a multitude - a multitude - of orphan genes. Of the 38,852 protein-coding genes in the ash tree one quarter (9,604) are found only in the ash.[38]

Unique and personal genes are sprinkled liberally throughout the plant and animal kingdoms, like peanut bits on sundaes, because they are what make each genus or species unique. Geneticists should have expected to find them. After all, something has to make a chicken special. At the same time, maybe God wanted to make His mark clear. Stamp.

Now! There are researchers working hard to track these things. The science of systematics is focused on determining close relationships between species and genera, families and classes. They look for parsimony between groups. That is, they work to determine which groups have the fewest evolutionary steps from one to another. Characteristics are lost and others gained, but the smallest number of steps is considered the most parsimonious.

The problem is that those steps can get pretty convoluted. The systematics experts are able to argue steps that are the most probable, but it's common for three different systematists to come up with

---

37 Baalsrud H.T. et al. (2018). De Novo Gene Evolution of Antifreeze Glycoproteins in Codfishes Revealed by Whole Genome Sequence Data. *Molecular Biology and Evolution*, 35(3):593-606. doi: 10.1093/molbev/msx311.
38 Sollars, E.S.A., et al. (2017) Genome sequence and genetic diversity of European ash trees. *Nature* 541:212–216.

three different sets of results. The variability and complexity in classifying organisms, the different methods used can end in different evolutionary conclusions. And fitting orphan genes into the mix adds an especially complicated layer of difficulty. It's all a huge topic.

Some Predictions

Okay! So, how do we handle all of this? It's no good to complain that we have no transitional forms to study and leave it at that. It's no good to whine like a chump and say, "We don't have in-between forms!" It's our job to make predictions and formulate ways to test those predictions. Let's start with making predictions about orphan genes.

Evolutionary biologists approach orphan genes with the assumption that they are recent developments. They compare the genomes of presumed near-relatives in order to single out the orphans in the mix. When they don't find homologs of human genes in chimpanzees, those orphans are regarded as genes that have developed since we split from the last common ancestors of humans and apes.

What tests could we run and predictions could we make to either confirm or discredit the *de novo* creation of orphan genes? How do we decide if common ancestry is real?

I want to try a few things. Just spitballing here:

1. We should hunt for "orphan" genes in a wide variety of creatures, not just assumed close relatives.

Orphans are clearly specific to each species and contribute to what makes that species unique. However, I wonder if some orphan genes are found in species not even remotely related to each other. That is, if DNA similarity is the result of a Programmer and not an evolutionary relationship, we might see orphan genes from one species show up in other completely unexpected species, with no apparent rhyme or reason.

It's not necessary, but I want to predict it. I predict we'll find homologs of the same orphan genes in species not closely related.

2. I predict that those multipurpose codes that sometimes behave as an intron and sometimes as an exon will prove far more complicated than anybody ever realized, and introns will be rejected as a playground where new genes can grow, because the code has significant purposes after all.

3. I predict that new genes introduced to a genome will only ever work if the regulatory genes are included as well. I predict the cell has no mechanism in place to adapt new genes spliced into a genome without accompanying instructions.

4. I predict that efforts to map out a step-by-step path backward from mice and humans to a common ancestor for proteins like rhodopsin will prove convoluted and painful, and the in-between forms will not function for anybody.

Seriously. I picture God just having fun. "Hey angels! Don't you like these octopus babies? Aren't they're so cool? They should have amazing eyes, but even though they're colorblind, let's make them able to change color and blend into their surroundings. Haha!! And check this out. I'm going to give the platypus a duck's bill and have it lay eggs. And I'm going to give koalas fingerprints like humans. But only koalas and humans, nobody else. And all kinds of animals should have prehensile tails, because hanging from one's tail would be so fun. Don't you think it would be a riot to swing from one's tail? Let's see… um… mice… and porcupines… opossums … and some monkeys and anteaters. And chameleons and sea horses and the skink! The skink. Hahah.. who named that thing, anyway? Enjoy yourselves, little creatures. Swing away, all of you!"

I recognize that God is grander than that, but I do think He enjoyed making everything.

Orphan genes might demonstrate how quickly genes can evolve, that they can regularly appear *de novo*, newly from nothing, and the cell is able to adopt them into its processes. Or, they might just be wrenches thrown into the mix to show that microbes-to-man evolution doesn't work.

I cannot possibly do justice to these issues in an appendix article, but I refuse to jump to the conclusion that evolution magically builds things. I declare that there be feasibility questions to test! All those of you testing the feasibility of these things, thank you!

So many questions. So many. What is the answer to it all?
Bryozoans!
Ha. Not really.
Still, I can stop being a jerk and finally introduce you to them.

Appendix

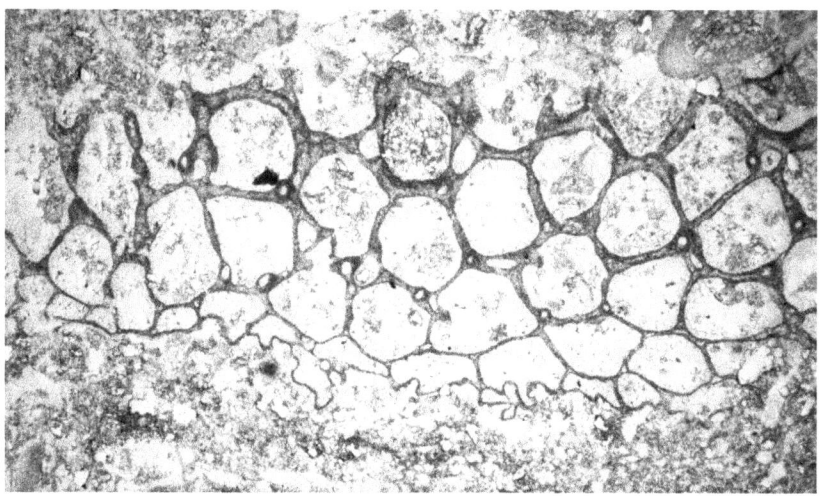

Figure 34: A trepostome bryozoan colony from the Toroweap Formation. Each "window" represents a chamber where a zooid lived.

6) What Are Bryozoans, Anyway?

As I write these words, Dr. Paul Denali Stillwell still professes to be an atheist. I believe God has something amazing planned for the old geologist, but I haven't lived to the end of his story yet. At this point in our lives, the most long-lasting result of our friendship is that he got me sucked into studying bryozoans.

It's time to introduce you to these odd creatures. I recognize that I've repeatedly mentioned them, but then I rudely - so rudely - never explain what a bryozoan *is*. Shoot, I'd never heard of bryozoans myself until Dr. Stillwell said, "Hey. You should study bryozoans with me!"

He tempted me. He lassoed me and pulled me in, and that's why I'm studying these crazy little aquatic organisms locked away in rocks.

First off, I like the word "bryozoan." It's a nice, friendly word. It's the sort of word I'd invite into my home to drink cocoa and snuggle on the couch to read Jane Austen. It's a warm, good word. Imagine if I had started studying the desmatosuchus. That's not a nice word at all! It sounds spikey and toothy, just like the crocodillian creature it names.

Figure 35: The polished surface of a piece of marine rock from the Toroweap Formation.

Figure 36: An acetate peel melted to the polished rock surface.

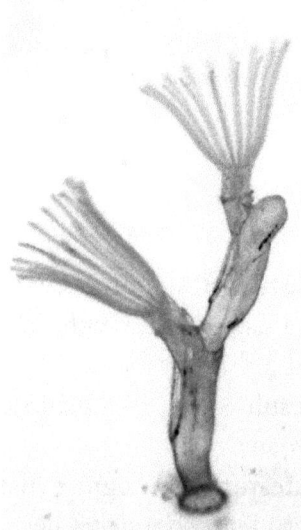
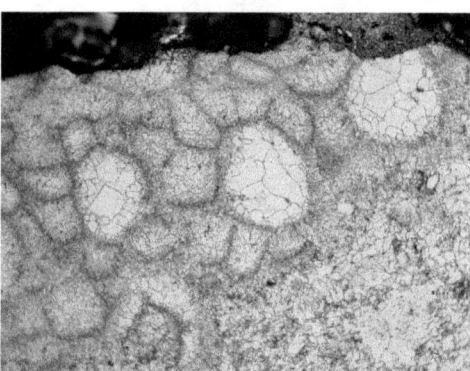

Figure 37: Left: a new bryozoan colony with two mature bryozoan zooids and a third beginning to bud. The lophophores are extended. Above: cystoporate bryozoan colony from the Toroweap. Below: cryptostome bryozoan colony from the Toroweap offering a longitudinal view of zooid chambers.

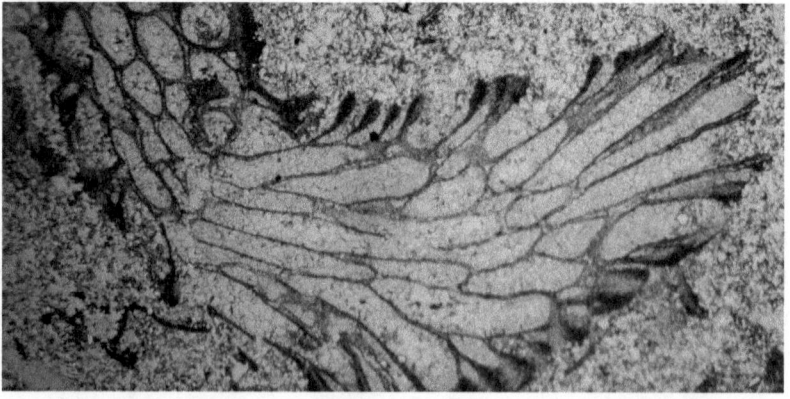

Appendix

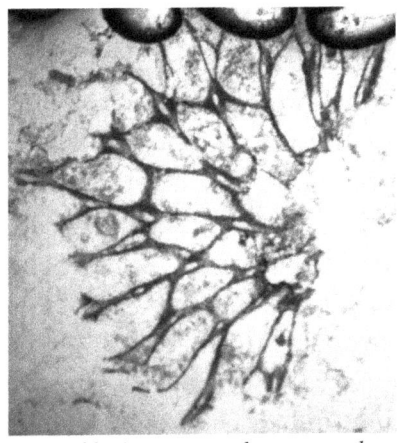

Figure 38: A trepostome bryozoan colony from the Toroweap Formation. I think she looks like a butterfly.

The word *bryozoan* derives from the Greek words for "moss" and "life" and basically means "moss animal." Bryozoans have absolutely nothing to do with moss, but they do have a habit of encrusting rocks or shells or other bryozoans with their pretty, lacey colonies. Dr. Stillwell is one of the world's few experts on Paleozoic bryozoans, and he made his name describing the fenestrate variety, which look like underwater ferns. The picture at the end of this section is from the chunk of bryozoans I studied in Dr. Stillwell's lab that summer of 2011. It's full of bryozoan fragments all piled together and on top of each other. Some bryozoans look like lace, and others look like underwater cheese puffs or sesame snacks. I think they all look delicious. In that picture, the bryozoans that resemble fishnets (or Ruffles potato chips) are fenestrate bryozoans.

Bryozoans are not plants, by the way. They are colonies of little animals, little zooids, and they are so distinct from all other life on earth that they have their very own phylum. Bryozoans aren't clearly related to anything else, like corals or sponges. They have their very own category in the animal kingdom.

But, what are bryozoans? They are multicellular, generally sessile aquatic colonial coelomates that are lophophore feeders with a u-shaped digestive tract. I'm kidding; I won't do that to you.

Bryozoans are cool, that's what they are. They live in the water and look a bit like coral, but they are *so* much spiffier than coral. They build colonial apartment buildings together, which are what get preserved in rock. Each of the holes in the lacey structure acts as a window into the chamber where a zooid lives. These zooids have guts, and their digestive tract is u-shaped, which means their anus is a different hole than their mouths, which is very civilized of them.

It's not far from their mouths, but that's okay because they don't want their waste to build up inside their little bedrooms.

Imagine a bryozoan larva swimming through the water, feeling exposed in the great, vast ocean. It finds a nice spot with a good view and settles down to metamorphose into an adult and start a new colony. It buds asexually to produce other sister bryozoans, like pussy willow babies up a stem, and together these sister zooids build themselves an underwater condominium out of calcium carbonate or chitin. Each zooid has her own little unit in the condo with a window that looks out on the world. She can reach out with her tentacled lophophore feeding tube and snack on passing food particulates, and then hide back inside to keep safe from passing shrimp and other carnivores.

"And the amazing thing," Dr. Stillwell told me, "is that the whole colony works together like a single organism, and nobody knows quite how they do it." It's like they're all wired together through the walls of their little bedrooms, and if one side of the colony is startled, all the lophophores pull inside at the same time.

Bryozoans are found all over the world. They are described in the geological record as early as the Ordovician, and they continue to bless coastal shelves to this day. Crazy Ernie agrees that bryozoans could be used as index fossils if they were described more completely, because they are both widespread and diverse. (Index fossils help geologists recognize which layer they've drilled into.) I think that bryozoans are *so* much cuter than conodonts (a typical index fossil). I mean really. Who wants to study nasty little eel-like jawless fish when they can study marine colonies that look like butterflies?

It doesn't seem like a hard choice to me.

For my Master's research, I slice open the rocks and take pictures of the colonies inside. I have to cut the rock, polish it flat, and dip it briefly in acid. The acid eats away the limestone rock just a tiny bit, so that the bryozoan fossils stick up above the surface. I have pictures of a polished rock on the previous page. See the little… circles and blobs in my polished rock? Those are bryozoan colonies.

I use acetone to melt an acetate "peel" to the polished surface

Appendix

of the rock. After a few minutes, I can pop the peel off and put it under a microscope and – boom – look at the top layer of that rock all magnified. I then hunt to find the remnants of a multitude of bryozoan colonies and take their pictures. Cheese puffs.

So! There you go! These are bryozoans. These are the critters that I studied with Dr. Stillwell the summer of 2011 - and ever after.

I now have a lab in my basement with a rock saw, a polisher, and a microscope with a camera connected to my computer. I have my very own little bryozoan-study lab all set up in my own house, because Dr. Stillwell thought I'd like to do research on these odd critters that few people know exist.

And I don't know if they smell like cilantro or not.

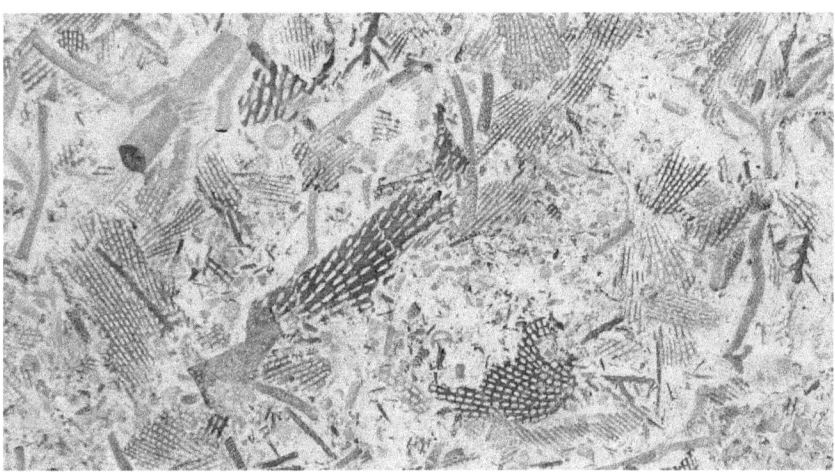

Figure 39: Mississippian rock from Illinois filled with bryozoan fossil fragments.

255

Books in the Science & Wonders Series

Volume 1: On The Edge of the Chasm

Volume 2: The Light, The Heat

Volume 3: As X Goes to Infinity

Volumes 4 and 5 are anticipated when the full story of Dr. Stillwell comes to completion. As of this printing, the author doesn't know how it ends.

Also by Amy Joy Hess

Gun Shot Witness: The Tim Remington Story
The true story of Tim Remington, the Coeur d'Alene pastor shot six times with .45 caliber hollow point bullets - by a man who believed Tim was an alien from Mars. Tim survived and returned to his church full of people God had rescued from the darkest of places.

Amy Joy and children at one of the world's oceans in the spring of 2011. They have returned since. Both to the ocean and to other undisclosed locations.

www.ingramcontent.com/pod-product-compliance
Lightning Source LLC
Chambersburg PA
CBHW070638050426
42451CB00008B/209